IFIP Series on Computer Graphics

Editors
J. L. Encarnação
G. G. Grinstein

G. Grinstein   H. Levkowitz (Eds.)

Perceptual Issues in Visualization

With 72 Figures, 17 in Color

Springer

Editors

Dr. Georges Grinstein
Dr. Haim Levkowitz
Department of Computer Science
Institute for Visualization and Perception Research
University of Massachusetts Lowell
One University Avenue
Lowell, MA 01854, USA

ISBN-13: 978-3-642-79059-1   e-ISBN-13: 978-3-642-79057-7

DOI:10.1007/ 978-3-642-79057-7

CIP data applied for

© IFIP Series on Computer Graphics 1995
Softcover reprint of the hardcover 1st edition 1995

Typesetting: Camera ready by authors with Springer TeX macro package
Cover design: Künkel + Lopka Werbeagentur
SPIN 10086715   33/3142 – 5 4 3 2 1 0 – Printed on acid-free paper

Preface

With the increase in the amount and dimensionality of scientific data collected, new approaches to the design of displays of such data have become essential. The designers of visual and auditory displays of scientific data seek to harness perceptual processes for data exploration. The general aim is to provide ways for raw data, and the statistical and mathematical structures they comprise, to "speak for themselves" and, thereby, enable scientists to conduct exploratory, in addition to confirmatory analyses of their data.

The present primary approach via visualization depends mainly on coding data as positions of visually distinguishable elements in a two- or three- dimensional euclidean space, e.g., as discrete points comprising clusters in scatter-plot displays and as patches comprising the hills and valleys of statistical surfaces. These displays are immensely effective because the data are in a form that evokes natural perceptual processing of the data into impressions of the presence and spatial disposition of apparent materials, objects, and structures in the viewers apparent physical environment. The problem with this mode of display, however, is that its perceptual potency is largely exhausted at dimension three, while we increasingly face the need to explore data of much greater dimensionality.

The challenge posed for visualization researchers is to develop new modes of display that can push the dimensionality of data displays higher while retaining the kind of perceptual potency needed for data exploration.

On October 23 and 24, 1993, during the IEEE Visualization '93 conference in San Jose California, an IFIP workshop was held to explore and discuss Perceptual Issues in Visualization. This was the first workshop of its kind, and its purpose was to provide, under the auspices of IFIP WG 5.10, an international forum to present, discuss, and evaluate new designs of displays for the visual and auditory exploration of scientific data, so as to make such displays more perceptually effective. The workshop brought together an international group of developers and users of both visual and auditory displays of data, as well as perception researchers, and provided ample time for discussions on a wide range of issues.

One obvious line of attack, and one focus of the workshop, is to raise the dimensionality of the current, conventional position-coding displays by giving the plotted elements additional discriminable features besides their position. The plotted elements can be given additional visible features that, like their positioning, could also be put under data control. Coding of the plotted elements using color, size, and shape have often been employed in attempts to raise the effective dimensionality of scatter-plot displays. Real color coding is now commonly employed, and with considerable perceptual effectiveness, to raise the dimensionality of surface displays. Various kinds of geometric codes (icons and glyphs) have also been tried. By providing ways for analysts to "pluck" or "strum" the visual elements electronically, their dimensionality can also be raised with data-controllable auditory features.

The papers submitted in response to the Call For Participation were grouped into three subgroups: User Interaction, Alternative Paradigms, and Test Data Sets. Each subgroup was tasked to describe the key research issues they believed are critical. This book contains a selected subset of the original papers submitted by the participants, as well as the three subgroup reports.

These proceedings offer an excellent reference to current research in this relatively new field of applied perception, and provide a timely portrayal of the problems that can be addressed now and in the future towards increasing the effectiveness of information displays. We hope that the readers will find a wealth of stimulating information in this book, and look forward to the research that it will stimulate.

The Editors

Georges Grinstein is Professor of Computer Science at the University of Massachusetts Lowell, in Lowell, MA, and Director of the *Institute for Visualization and Perception Research*. He is also a Principal Engineer with the MITRE Corporation. He received his B.S. from the City College of N.Y. in 1967, his M.S. from the Courant Institute of Mathematical Sciences of New York University in 1969 and his Ph.D. in Mathematics from the University of Rochester in 1978. Dr. Grinstein is a member of IEEE's Technical Committee on Computer Graphics and on the editorial board of several journals including Computers and Graphics and the Eurographics Society's Computer Graphics Forum. He was vice-chair of the executive board of IFIP WG 5.10 (Computer Graphics) and was co-chair of the IFIP Conference on Experimental Workstations held in Boston in 1989. He was panels co-chair for Visualizations '90 (San Francisco), program co-chair for Visualizations '91 (San Diego), conference co-chair for Visualizations '92 (Boston) and for Visualizations '93 (San Jose), co-chair of the IFIP 1993 Workshop on Perceptual Issues in Visualization, and co-chair for the IEEE Workshop on Database and Visualization Issues. He has chaired several committees for the American National Standards Institute (ANSI) and the International Standards Organization (ISO). He is co-chair of the SPIE '95 Visual Data Exploration and Analysis Conference.

His areas of research include graphics, imaging, sonification, virtual environments, user interfaces and interaction, with a very strong interest in the visualization of complex systems.

Haim Levkowitz is Assistant Professor of Computer Science and a founding faculty member of the *Institute for Visualization and Perception Research* at the University of Massachusetts Lowell, in Lowell, MA. Since 1982, he has been studying multidimensional, multiparametric imaging and visualization. He has developed new color methods for computer graphics representation of parameter distributions and methods for evaluation. He is the developer of the Generalized Lightness, Hue, and Saturation (GLHS) family of color models, the Linearized Optimal Color Scale (LOCS), and the Color Icon. He has also developed and implemented tools for automated psychometric evaluation of the developed display methods, and has used these tools to conduct observer performance evaluations of some of his methods. Dr. Levkowitz is the author of over 25 publications on color, visualization, and imaging. His panel at Visualization '91, "Color vs. Black-and-White in Visualization" won the Best Panel award. He has organized and taught several tutorials at conferences on various topics in visualization. Dr. Levkowitz was on the conference committees of Visualization '91, '92, and '93; on the program committee of the 46th IS&T meeting; Co-chair of the 1992 Boston Workshop on Volume Visualization; and Co-Chair of the IFIP 93 Workshop on Perceptual Issues in Visualization.

Table of Contents

1 Introduction

The purpose of collecting information is to present it to, and have it analyzed by human beings. The amounts of information collected nowadays, and its complexity are constantly increasing. With such increases, the complexity of the analysis also increases, sometimes to the point where it is impossible to comprehend, let alone analyze the information. There is a growing need to understand the nature of presentations that make the comprehension and analysis of information easier and more efficient. With that knowledge and understanding, to achieve an optimal transfer of information, presentations of data have to be designed with the human audience in mind. The process depends on the audience to which information is presented, the goals of the presentation, and the questions to be answered. Human perceptual capabilities must be well understood and exploited to enhance, rather than hinder the presentation. Perceptually-based presentations harness human perceptual channels to pre-consciously segment raw information into meaningful groups and categories. Pre-conscious processes are parallel, hard-wired or entrained, fast, and relentless. They occur simultaneously with conscious analysis and thus do not interfere with our ability to think. On the other hand, conscious processes are serial and thus require scrutiny. They are ad-hoc, slow, and can cause fatigue, thus interfering with higher level analysis.

Examples of tasks that can benefit from such presentations include detecting a tumor in a medical image, finding a trend in collected population data, or segmenting text. Detecting a grayscale tumor in a collection of other grayscale blobs may need a conscious search; detecting a red tumor in a group of green blobs is pre-conscious, and will occur instantaneously irrespective of the number of blobs. Finding a specific word in a body of text requires a conscious search; the time it takes to find the word will depend on the number of words in the text. (The choice of the font type, size, and color of the sought word, as compared to those of the rest of the text may make the task easier or hard.)

Current visualizations are limited and limiting. Their blatant weakness is in one neglected area, the area of harnessing users' perceptual capabilities to accomplish the most effective visualizations. We refer to this harnessing as applied perception. Applied perception requires first a deep understanding of human perception in general. Once such a comprehensive understanding is achieved, it can be harnessed to accomplish optimal presentations of information. The areas of perception that can be harnessed include, in addition to the obvious visual, the auditory, the olfactory, the tactile, and potentially others. The focus of this workshop was limited to visual (mostly) and auditory perception, as they are applied to presentation of information.

Harnessing human perception can increase in a number of ways not only the effectiveness of the presentation but also the amount of data explored. It can increase the number of dimensions and the number of parameters that a user may interact with and explore effectively. It can bring out structure in data hereto unwitnessed with the limited visualizations of today. Current visualizations typically display three parameters, in a spatial representation (mostly two-dimensional, though occasionally three-dimensional), with animation capable of representing an additional dimension, usually time.

Considering the types and amounts of data that can be collected nowadays, this is quite limiting. NASA's Earth Observing System (EOS[3]) is expected to generate over a terabyte of data per day (this is more than the entire amount of data collected by NASA throughout its existence). There are even some discussion of acquisition of petabyte databases. Such large datasets cannot be analyzed using the traditional ways. They will require new browsing techniques employing rapid presentations of the data in ways that will allow users to explore and capture important structures and relationships in the information within very short periods of time. This will require a profound understanding, and utilization of preattentive perception of the visual system (and other sensory systems) to make such presentations salient.

The visual system by itself employs several layered perceptual mechanisms. The first, and strongest perceptual mechanisms are basic discrimination in color and line orientation. Indeed, we are very quick to detect a blob of different color from other blobs. We are also able to very quickly detect a line of different orientation amongst lines of similar orientations. Further layers of visual perception provide for detection of, and discrimination among more sophisticated elements, which may differ in their shape, size, orientation, texture, the number of terminals, or any combination thereof.

Each one of these perceptual mechanisms provides a potential carrier of coded information, but not all of them provide the same detection and discrimination capabilities. For example, while color is very powerful in general, the use of colors that differ from each other only in hue, but not in lightness, will make discrimination amongst them (and thus amongst the data items they code) more difficult, at times impossible. A simple change in the selection of color mappings can make the difference between instantaneous discrimination and no discrimination at all. The same holds true for codings that utilize textures, orientation, or shape; very similar considerations apply to auditory, tactile, and other senses.

The purposes of this workshop were to draw from the vast knowledge and experience that is available in the basic perceptual research community, and to apply it to the specific problems that the visualization community is facing. In particular:

1. To bring together visualization researchers and basic perception researchers, and to start a dialog between these two groups.
2. For visualization researchers to acquire an understanding of human perceptual mechanisms from the perception researchers.
3. For perception researchers to acquire an understanding of the problems that visualization researchers are facing, with the hope that such an understanding would steer their own research in directions that might help solve such problems.

---

[3] EOS Reference Handbook, 1993, Ghassem Asrar and David John Dokken, (Editors), Earth Science Support Office, Document Resource Facility, 300 D Street, SW, Suite 840, Washington, DC 20024

The workshop accomplished item number 1. Researchers from both communities got together and established the beginning of the desired dialog. Items number 2 and 3 are of a much longer term nature. It is only through the continuation of such dialogs, and through establishing collaborative projects between the two group, that these goals will be achieved.

1.1 Workshop Format and Goals

The workshop provided opportunity for the presenters to discuss their positions as well as to participate in several group discussions.

The first group focus of the workshop dealt with perception issues. Eliot Handelman of Princeton University discussed the need for, and the design issues in building an auditory-tactile cyberspace. He highlighted the importance of moving sound to the forefront of the display environment. Haim Levkowitz of the University of Massachusetts Lowell discussed the similarities and differences between vision and sound perception. He brought forth the importance of synergies between the two forms of data presentations, and demonstrated how vision and sound can support or hinder each other as mechanisms for information presentation. Ronald M. Pickett, also of the University of Massachusetts Lowell, discussed issues in harnessing preattentive processes in visualization, and identified key perceptual elements that can be harnessed in the presentation of data. Penny Rheingans, of the US EPA Visualization Center, discussed some fundamental perceptual principles for effective visualizations, and emphasized the importance of color control for such effective displays. Bernice E. Rogowitz from IBM Research highlighted the importance of integrating knowledge/expert systems with presentation systems. She described two rule-based visualization systems harnessing principles of human perception. Finally, Tom Zier from Beloit College discussed presence in the binocular field and highlighted the importance of looking at alternative presentation models.

The second group focus of the workshop explored the mathematical, statistical, and computational issues that must be dealt with in the exploration of alternative representations of data. Lack of validation was pointed out as a key factor in the lack of acceptance of these methods. William Hibbard from the University of Wisconsin Madison discussed interaction with high dimensional data displays. He brought forth the importance for users to interact with displays of dimensionality larger than three, and described a mechanism to accomplish this. Philip K. Robertson of CSIRO discussed Markov random field textures and perceptual control issues. Pak Wong and R. Daniel Bergeron from the University of New Hampshire discussed the importance of generating test data in order to compare the relative effectiveness of visualization techniques and approaches. They described a multidimensional multivariate image evaluation tool that permits some control over the generation of test data. Finally, Janet Haswell from Appleton Rutherford Laboratories discussed issues, computational in nature, in the visualization of unsteady flow data.

The third and final group focus of the workshop was on the mapping issues between the data and its representations. Georges Grinstein and Suraiya Haque from the University of Massachusetts Lowell discussed the fundamental elements of icon generation with the goal of building performance evaluation systems for large families of geometric and color icons. Daniel Keim and Hans-Peter Kriegel, from the University of Munich discussed some of the issues in the visualization of large amounts of retrieved multidimensional data, and proposed some solutions. G. Deon Oosthuizen from Stanford University discussed visualization alternatives for specially structured representations of data with an example of a visualization of a lattice for the exploratory analysis of categorical data. Finally, Florian Schroeder from the Fraunhofer Institute for Computer Graphics, Darmstadt, discussed the importance of understanding audience differences and determining audience preferences in the visualization of meteorological data.

The workshop participants then broke into these three groups to discuss and develop grand challenge questions that would be presented to the whole group for further discussion. These working groups met several times and generated reports summarizing their discussions.

The next section describes the contents of these proceedings.

1.2 Contents Overview

The three main subgroup reports are provided first, followed by a selection of those papers that were submitted for final publication.

Workshop Subgroup Reports—Research Areas In Chapter 2, Keim, Bergeron, and Pickett's report, "Data Sets for Evaluating Data Visualization Techniques," takes a step toward addressing the problem how to measure the effectiveness of visualization systems. The step the authors take is to define a model for specifying the generation of test data that can be used for standardized and quantitative testing of a system's performance. These test data sets, in conjunction with appropriate testing procedures, can provide a basis for certifying the effectiveness of a visualization system and for conducting comparative studies to steer system development.

In Chapter 3, Hibbard, Levkowitz, Haswell, Rheingans, and Schroeder's report "Interaction in Perceptually-Based Visualization," addresses the tight relationship between good interaction and perceptually-based visualization. They show the importance of interaction in the development of visual maturity in animals and human beings, and describe various essential aspects of interaction, as applied to accomplishing better perception of visualized data.

Papers In Chapter 4, "Harnessing Preattentive Perceptual Processes in Visualization," Pickett, Grinstein, Levkowitz, and Smith present an overview of thier visualization work at the University of Massachusetts Lowell. They describe their general approach of creating iconographic displays and the perceptual rationale for how they design those displays. They then describe their accomplishments to date, mainly developing a general purpose system for creating iconographic displays and illustrating displays with different types of icons and on a wide variety of databases. They conclude with a description of their main long-term goals, to develop a more capable display system and to conduct basic applied research.

In Chapter 5, "Environment and Studies for Exploring Auditory Representations of Multidimensional Data," Levkowitz, Pickett, Smith, and Torpey describe the characteristics of a workstation they have developed, with the capability to run quick psychometric tests to obtain quantitative figures of merit for alternative auditory representations. This is a requirement for auditory-display researchers engaged in the development of new technologies. The authors also describe a testing methodology they propose for the development of new auditory data displays of a type that they have been working with for the last few years. Finally, they describe a specific set of studies they are now conducting.

In Chapter 6, "Perceptual Principles for Effective Visualizations," Rheingans and Landreth argue that since visual data representations are perceived through the filter of the human visual system, it is imperative that the characteristics of this system be taken into account during the design and rendering of visual displays. Their paper presents a set of perceptual guidelines for the construction of effective visualizations. They demonstrate the impact of suggested techniques with side-by-side pictures .

In Chapter 7, "Interactivity and the Dimensionality of Data Displays," Hibbard, Dyer, and Paul use mathematical models of data and displays to illustrate the importance of distinguishing between independent and dependent variables when counting the dimensions of data sets and displays. The authors argue that the number of independent variables occurring as dimensions of a display model is the most important factor determining its information carrying capacity. They illustrate with their visAD system that independent variables in a display model require interactive techniques for their implementation. They conclude that interactivity is critical for visually communicating large amounts of information, and that the perceptual properties of interaction techniques are an important topic for visualization research.

In Chapter 8, "Towards Perceptual Control of Markov Random Field Textures," Li and Robertson propose a method to establish a perceptually meaningful Euclidean space for textures generated as samples of first-order Markov Random Fields (MRF). The authors show how, under a definition of texture difference, within a specified neighborhood of zero, the first-order MRF parameters can be considered as orthogonal, and approximately define textures in a uniform manner. They suggest that the Euclidean space established with this texture distance metric may be considered a perceptual texture space that can be used for visualization purposes. Their definition of texture difference is based on texture classification results that are consistent with human visual discrimination when experimented with a set of test images. They present also demonstrations of using their texture space as a preliminary visualization tool.

In Chapter 9, "A Multidimensional Multivariate Image Evaluation Tool," Wong and Bergeron describe their current research focus on the representation and visualization of multidimensional multivariate (mDmV) data. The authors are developing a visualization evaluation tool, whose primary goal is to provide an environment for visualization researchers to evaluate human responses to different computer generated visual images. The tool has the ability to create mDmV data with embedded stimuli, and display them in a variety of ways including icons. In addition, it provides statistical analysis functions for visualization researchers to study relationships among variates.

In Chapter 10, "Visualizing Electromagnetic Data," Haswell looks at how a commercial visualization system was used to produce interactive visualizations of simulated Electromagnetic data. The author describes the weaknesses in the system by looking at techniques available for viewing vectors on an unstructured grid. The user of the simulation package takes advantage of symmetry within the problem to reduce the amount of computation required. Rather than duplicate unstructured grid data, a simple mirror tool has been developed to duplicate geometries instead. Finally, the author descries how these techniques were used to perceive the changing flow in time-varying 3D vector field data.

In Chapter 11, "Visualizing Large Amounts of Multidimensional Data: Possibilities and Limits," Keim and Kriegel describe their concepts to visualize very large amounts of multidimensional data. Their visualization technique, which has been developed to support querying of large scientific databases, is designed to visualize as many data items as possible on current display devices. Even if they are able to use each pixel of the display device to visualize one data item, the number of data items that can be visualized is quite limited. Therefore, in their system they introduce reference points (or regions) in multidimensional space, and consider only those data items that are "close" to the reference point. The data items are arranged according to their distance from the reference point. Multiple windows are used for the different dimensions of the data. The distance of each of the dimensions from the reference point (or region) is represented by color. In exploring the database, the reference point (or region) may be changed interactively, allowing different portions of the database to be visualized. To visualize larger portions of the database, sequences of visualizations may be generated automatically by moving the reference point along some path in multidimensional space. Besides describing the authors' visualization technique and several alternatives, they discuss some of the perceptual issues that arise in connection with our visualization technique.

In Chapter 12, "Using a Lattice for Visual Analysis of Categorical Data," Oosthuizen and Venter address the problem of visualizing data of dimensions larger than three. The method the authors describe is primarily aimed at so-called categorical (or nominal) data, but also applies to other discrete-valued data, such as, data entities defined in terms of vectors of discrete attribute values. Categorical data is data with unordered attribute values, mostly because they are symbolic. Data with boolean values constitute a special case. The authors method utilizes the fact that data entities can be regarded as points in an n-dimensional space, where n is the number of attributes recorded for each data entity. Although it is not possible to depict an n-dimensional cube using three-dimensional graphics, it is possible to provide the user with an abstraction of the structure of the data by revealing the non-empty subspaces and how they are related. The graphical user interface allows the user to browse through the space of abstractions.

In Chapter 13, "Audience Dependence of Meteorological Data Visualization," Schroeder argues that in order to visualize the data in a perceptually effective way, the audience who will view the results must be considered. The importance of audience dependence varies among different fields of applications of visualization techniques. The author discusses the importance of audience dependence in visualization of meteorological data. Expected audiences plays a very important role in determining the way data is converted into images. The author describes a system he has developed for visualizing weather-related data for meteorological researchers as well as for the general audience of television watchers. Each of these groups has its own specific demands. Meteorologists need a presentation containing their symbols and possibly many data sets at the same time to get a better understanding of their data and simulation models. The general audience need images or sequences they can understand intuitively and easily. Schroeder and his colleagues categorized the types of meteorological data, and determined the needs of both groups. For example, to visualize cloud-specific weather data for the general audience, the developers incorporated fractal functions to show realistic images of clouds. The system described here is used both by meteorological researcher and for the daily production of weather forecasts for several German television stations.

Acknowledgements

We would like to thank Paul Breen and the MITRE Corporation of Bedford, Massachusetts, who provided logistical support for the workshop. We are also indebted to the committee members who helped make this workshop, and these proceedings happen.

International Program Committee

Co-chairs

Georges Grinstein — University of Massachusetts Lowell and the MITRE Corporation
Haim Levkowitz — University of Massachusetts Lowell

Program co-chairs

Ron Pickett — University of Massachusetts Lowell
Stu Smith — University of Massachusetts Lowell

Program Committee

- Pere Brunet, Dept. de Llenguatges i Sistemes Informatics, Barcelona, SPAIN,
  brunet@lsi.upc.es
- Steve Cunningham, California State University, Stanislaus, Turlock, CA,
  rsc@csustan.edu
- Jose Encarnacao, ZGDV Darmstadt Germany,
  jle@igd.fhg.de
- Bianca Falcidieno, IMA CNR, Genova, Italy,
  falcidieno@image.ge.cnr.it
- Martin Fruehauf, ZGDV Darmstadt, Germany,
  fruehauf@igd.fhg.de
- Georges Grinstein, Institute for Visualization and Perception Research, University of Massachusetts Lowell, Lowell, MA, USA, and The MITRE Corporation Bedford MA,
  grinstein@cs.uml.edu
- Markus Gross, ZGDV Darmstadt, Germany,
  gross@igd.fhg.de
- F. R. A. Hopgood, Rutherford Appleton Laboratory, UNITED KINGDOM,
  frah@ib.rl.ac.uk
- Christoph Hornung, Fraunhofer-IGD Darmstadt, Germany,
  hornung@igd.fhg.de
- Haim Levkowitz, Institute for Visualization and Perception Research, University of Massachusetts Lowell, Lowell, MA, USA,
  haim@cs.uml.edu
- Leo Pini Magalhaes, UNICAMP-FEE-DCA, Campinas—SP BRAZIL,
  leopini@bruc.bitnet
- Thomas Papathomas, Rutgers University, Rutgers, NJ, USA,
  papathom@gandalf.rutgers.edu
- Ronald Pickett, Institute for Visualization and Perception Research, University of Massachusetts Lowell, Lowell, MA, USA,
  pickett@cs.uml.edu
- Philip K. Robertson, CSIRO Division of Information Technology, Canberra, Australia,
  phil@csis.dit.csiro.au
- Bernice E. Rogowitz, Exploratory Visualization Group, IBM T.J. Watson Research Center, Yorktown Heights, NY, USA,
  rogowtz@watson.ibm.com
- Stuart Smith, Institute for Visualization and Perception Research, University of Massachusetts Lowell, Lowell, MA, USA,
  stu@cs.uml.edu
- Jose Carlos Teixeira, Departamento de Matematica, Universidade de Coimbra, Coimbra, PORTUGAL,
  teixeira@matuc2.mat.uc.pt

Test Data Sets for
Evaluating Data Visualization Techniques

Daniel A. Keim
Institute for Computer Science, University of Munich,
Leopoldstr. 11B, 80802 Munich, Germany,
keim@informatik.uni-muenchen.de

R. Daniel Bergeron
Department of Computer Science,
University of New Hampshire,
Durham, NH 03824, USA, rdb@cs.unh.edu

Ronald M. Pickett
Institute for Visualization and Perception Research,
University of Massachusetts Lowell,
Lowell, MA 01854, USA, pickett@cs.uml.edu

Abstract. In this paper we take a step toward addressing a pressing general problem in the development of data visualization systems — how to measure their effectiveness. The step we take is to define a model for specifying the generation of test data that can be employed for standardized and quantitative testing of a system's performance. These test data sets, in conjunction with appropriate testing procedures, can provide a basis for certifying the effectiveness of a visualization system and for conducting comparative studies to steer system development.

Keywords: Testing Data Visualizations, Generating Test Data, Visualizing Multidimensional and Multivariate Data, Perception of Visualizations

1. Introduction

Data visualization has captured very high interest among scientists and many commercial and public domain visualization systems have appeared in recent years including, for example, *AVS* [Ups 89], IBM's *Data Explorer*, Silicon Graphics' *Explorer*, *PV-Wave* from Precision Visuals, *IDL* from Research Systems, *Khoros* from the University of New Mexico, and *apE* from Ohio State [Dye 90]. All generally available visualization systems rely on conventional visualization techniques based primarily on two-dimensional displays, or two-dimensional surfaces in a three-dimensional world.

Considerable efforts have also been aimed at developing and prototyping non-traditional visualization techniques that attempt to present multivariate and multidimensional data in effective ways. Many different approaches have been demonstrated, and their potential value in numerous areas of application have been touted. Some examples include work by Grinstein et al. [PG 88, GPW 89, SBG 91], Beddow [Bed 90], LeBlanc et al. [LWW 90], Inselberg and Dimsdale [ID 90], Beshers and Feiner [BF 92, BF 93], Keim et al. [KKS 93, KK 94], and Mihalisin [Mih 90]. Lacking in all this activity is any quantitative evidence of just how effective the techniques are. Until we develop a basis for evaluation, we will not be able to get beyond this current demonstrational stage. To progress, we need to know with certainty what is working and what adjustments are leading to improvement.

The general purpose of a visualization system is to transform numerical data of one kind or another into pictures in which structures of interest in the data become perceptually apparent. By encoding and formatting the data into just the right kind of pictorial array, the structures, so the hope goes, will make themselves perceptually apparent. Conceivably, one might find a kind of coding and formatting that reveals many different kinds of structures in the data. But it is also conceivable that some structures might require very narrowly tuned codings and formats to become perceptible.

One of the big weaknesses of our present state of understanding is that we hardly know what we mean by *structure in the data*. We know something about, and even have a precise language for describing, some familiar and simple statistical structures. We turn to this familiar domain of structures for the test data sets proposed in this paper. But the field is in great need of a broader conception and language of structure. We need a taxonomy to inventory the world of structures that visualization systems might need to address. Creating an awareness of this lack of understanding is, indeed, one of the ancillary goals of this paper.

Visualization systems are actually just another instance of technology in science for detecting, analyzing and interpreting signals —albeit signals (what we are calling structures) of a rather broad and often ill-defined type. The need to provide a basis for quantitative evaluation of systems for signal detection and recognition is well recognized in many areas of science and technology. Evaluation of medical diagnostic systems provides a good case in point. Medical imaging systems are subject to various objective certification tests with standardized "phantom" images to verify that they can reveal the details of images that have to be resolvable for certain types of diagnoses. Even beyond such general certifications are standardized evaluations to determine how well the whole system, including the radiologist who does the reading of the images, performs in detecting and diagnosing particular diseases. In those situations, sets of test patterns (images of real cases) are assembled, and standardized tests are conducted to measure exactly how well the system performs (see [SP 82]). The need for, and approach to, evaluating visualization systems is almost exactly analogous. We want to know how well a given visualization system does in helping a scientist to detect and interpret a structure in his or her data. We need a standard set of test data sets and a standardized testing procedure. In this paper, we provide a start toward building this needed resource.

Our goal is to generate test data sets with characteristics similar to those of real data sets. Unlike real data sets, however, the characteristics of artificially generated data sets may be varied arbitrarily. We may, for example, vary the correlation coefficient of two dimensions, the mean and variance of some of the dimensions, the location, size and shape of clusters, etc. Varying the data characteristics in a controlled manner is crucial for evaluating different visualization techniques. For example, controlled test series allow us to find the point where data characteristics are perceivable for the first time, or the point where they are no longer perceivable. Also, the same test data may be used in comparing different visualization techniques, helping to determine their strengths and weaknesses.

2. Scientific Data

We are interested in generating test data that have characteristics similar to those of typical scientific data. Scientific data is characterized by its *data type*, the way in which it is *organized*, and the way in which the values relate to each other (the *distribution*).

2.1 Data Types

Scientific data can consist of multiple values of various data types, which are typically described using terminology from programming languages, such as *float*, *integer*, and *string*. For our purposes we are more interested in the generic characteristics of the data types. These are best identified using terminology from the statistical domain, which defines the following standard types:

nominal — data whose values have no inherent ordering

ordinal — data whose values are ordered, but for which no meaningful distance metric exists

metric — data which has a meaningful distance metric between any two values.

2.2 Organization of the Data

Scientific data is often highly organized in that data values have some inherent physical or logical relationship to other data values, which might be called its *neighbors*. This organization is usually called its *data structure*. Note the distinction between the data structure (the structure *of* the data) and the patterns of values in the data that we are trying to see with a given visualization technique (the structures *in* the data). We are primarily interested in scientific data that is organized with only limited inherent structure —in particular, we consider here only data that can be represented in arrays. This restriction omits engineering-style data that is most naturally represented with more complex data structures.

The least-structured form of data is a set of records which have no particular *a priori* ordering among themselves. Conventional database records satisfy this requirement. Although there may be many fields in the records that *could* be used to order the records, there is no pre-defined ordering that is inherent in the data. Database keys which are used to uniquely identify and access database records, also do not provide a natural ordering since in most cases they only induce an artificial ordering of the records. Data sets having no inherent structure or organization can be considered to be 0-dimensional arrays.

Other data has underlying structure or organization, such that each data record has an inherent unique *position* relative to the other records. Often the record's position is related to a location in some geometric domain, or to a point in time. Such data can be generated by sampling of physical phenomena or from simulations and is commonly represented as arrays (perhaps multidimensional). A record can now be identified and accessed by its relative position in the data set which corresponds to the indices into its position in the (multidimensional) array. If this position is determined by the coordinate values of its placement in the geometric and/or time domain, these coordinate values are likely to be explicitly included in the data record. However, if the data elements are uniformly distributed over the range of indices of the array, their values can be computed from the indices of the record into the array, and need not be explicitly stored. If a data variable maps to an index into the data set's storage array, we say that that variable represents an *array dimension* of the data set.

Regardless of how the data is initially defined, the visualization may choose whether or not to place a record's visual representation on the display in a way that is consistent with the record's position in the data set. For example, consider a data set composed of carbon and nitrogen measurements on a two-dimensional x-y grid. A straightforward visualization might show the carbon value as a color or intensity at each position on the x-y grid; i.e., the x-y grid of the data is mapped to the x-y coordinates of the display. However, it might also be useful to produce a visualization in which the values of the carbon and nitrogen are mapped to the x-y coordinates of the display and the y-value of the grid is mapped to the intensity. (Note that this mapping need not produce a single-valued function: there may be multiple y-values for one pair of carbon/nitrogen values. If the visualization technique must have only a single value, some choice has to be made.)

2.3 Examples for Typical Data Sets

Our model of the data and the data generation process allows us to handle a wide range of types of data in a uniform way. In the following, we provide examples for typical data sets that may be generated using our model.

Statistical Data

We use the term statistical data to describe data sets whose data values are best defined by statistical parameters such as distribution functions, correlation coefficients, or cluster descriptions. Statistical data may have an arbitrary number of dimensions with none of them being an array dimension. The data may be scattered arbitrarily in multidimensional space and, in general, even duplicate data items are allowed. Examples of this kind of data are financial databases, product databases, personal databases, databases that record banking transaction, telephone calls or other events, and scientific databases (e.g., observations or simulations). Most of these data sets are typically stored in relational database systems.

For evaluating different visualization techniques, it is interesting to study how well different visualization techniques represent statistical patterns described by some statistical parameter. For controlled studies of this type, the statistical parameters should first be varied one at a time. After understanding their effects on the visualizations, more realistic test data sets may be built by using multiple statistical parameters to describe the test data. Examples of data sets that are best described by distribution functions include deviations of norm values which are best approximated by normal distributions, radioactivity which may be described by an exponential distribution, or periodic events which may often be assumed to follow a uniform distribution. Single dimensions with such distribution characteristics may be specified easily.

If something about the relationship between multiple dimensions is known, the data may be better described by correlation coefficients and functional dependencies. The relationship of solar radiation and temperature, for example, may be described by a high correlation coefficient and some functional dependency. Since there are usually complex relationships between multiple dimensions in real data, we also provide the ability to specify correlations between multiple parameters and complex functional dependencies. An example of a more complex relationship is the interdependencies between temperature, humidity, solar radiation, precipitation and wind speed.

Local correlations are also important features of many data sets. In a local correlation the correlation coefficient is much higher in a specific region than in the whole data set. One way of describing this kind of relationship is to specify the different partitions of the data space separately. Another way of describing complex relationships is to consider them to be multidimensional clusters in an otherwise homogeneous, possibly empty multidimensional data space. Examples of data sets that can be best described by a base data set and a set of clusters are data sets that contain a portion of data items having some clearly distinguishable properties. We may also have time series of statistical data. In most cases, the time dimension is an array dimension. This means that the cardinality of the data set is given by the considered time frame and no duplicate data items may occur.

Image Data

Another important class of test data is image data. Image data is two-dimensional in nature. In terms of our test data generation, normal two-dimensional image data is generated by setting the total number of dimensions to 3 and the number of array dimensions to 2. Depending on the application, however, image data may have a much higher dimensionality since multiple values for each point of the two-dimensional array may occur or different types of images for the same region may exist. In earth observation science, for example, researchers record many images at different wavelengths. To specify the test data, first the ranges for the array dimensions need to be specified. The ranges of the array dimensions determine the total number of data items. Then, the specific characteristics of the data can be specified using distributions, functional relationships, (local) correlations or cluster descriptions. Note that only the characteristics of the non-array dimensions may be specified since the array dimensions are dense and their values are given by the range definitions. In many cases, however, the distributions, functional relationships, (local) correlations or cluster descriptions include some dependency on the array dimensions. We may further have time series of image data which requires a third array dimension.

Other Data

Image data may be easily extended to volume data by using an additional array dimension for the third dimension of the volume. Volume data and other types of data such as geographic, geometry, molecular, fluid dynamics or flow data have specific characteristics which can only be specified by our method to a very limited extend. For molecular data, we may, for example, generate a set of atoms and some random 3D structure. However, for such molecule data to be realistic, many physical, chemical, and biological constraints apply which have to be modeled explicitly. In general, generation of arbitrary realistic test data sets would require lengthy descriptions or complex simulations reflecting all constraints and interdependencies.

At this point, we want to stress that our goal is to test and compare visualization techniques for statistical and image data. We do not intend to produce test data sets that are completely realistic for some application domain. Instead, we want the test data sets to have only a few characteristics of real data sets. Important, however, is the possibility to vary the characteristics of the test data gradually. Although real data sets are very important in testing and comparing visualization techniques, we believe that an in-depth evaluation of their strengths and weaknesses is only possible with generated test data sets whose characteristics can be precisely controlled.

3. Structures in the Data

In order to generate large amounts of data, we need to have an automatic mechanism for generating the data with carefully controlled statistical variations. In some cases, we want to generate the values of a particular data field without regard to other neighboring values, or values of other fields; more often we want to model actual data that has some kind of *correlation* among the various data fields.

3.1 Probability Distributions, Correlations and Functional Dependencies

A test generation utility needs to support the ability to specify that data generation should be driven by a variety of probability distributions, including at least the well-known distributions such as poisson, gamma, gaussian, etc. [Dev 87]. These distributions require the user to specify parameters such as the maximum, minimum, mean, and standard deviation.

More complicated (and more realistic) data generation requires that the values of different fields in the data have some functional relationship to values of other fields. In fact, the quintessential goal of scientific data visualization is to assist the scientist in determining the nature of some phenomenon by understanding the relationships present among the data values that represent that phenomenon. By generating test data containing known relationships, we hope to be able to evaluate visualization techniques to see if these relationships produce a distinctive recognizable pattern in the visual presentation of the data.

The standard measure of correlation used in today's statistics packages is the *correlation coefficient* which is defined as a measure of the linear relationship between two variables. As useful as this measure is, it does not serve to identify more complicated relationships such as non-linear dependencies and dependencies based on 3 or more variables simultaneously. Since we are generating new data, rather than analyzing existing data, we can easily generalize the notion of correlation coefficients to specify more complex interrelationships. The basic mechanism for controlling the generation of interrelated data fields is to have the user define functional dependencies among these data fields. The *functional dependencies* allow the user to specify a formula to generate a set of *initial* values for a data record, but the user can also specify that a random perturbation should be applied to these values in order to approximate real data more realistically. The randomizing function parameters are under user control.

3.2 Data Clusters

Our model of a visualization evaluation environment is based on the notion that the test data set should contain subsets that have data characteristics which are distinctive from the rest of the data. The visualization test then presents the data (perhaps in a variety of formats) to see whether the distinctive subset produces a distinctive and recognizable visual effect. We use the term *data cluster* to refer to a subset of data with distinctive data characteristics. The specification of a data cluster requires the specification of a region of the data space as well as the data generation parameters to be used for generating data in that region.

In its most general form, a *region* is any contiguous subset of the n-dimensional data space defined by the set of fields in the data records of the data set. In its simplest form, we can define a *rectangular* region by identifying a specific range of data values of a subset of the fields. For example, a 2-dimensional rectangular region could be defined by specifying $23 \leq x \leq 45$ and $102 \leq y \leq 150$, for the fields x and y.

A precise definition of the notion of *distinctive data characteristics* is difficult to achieve and perhaps not even desirable. What constitutes significantly different data characteristics in one domain may not be significant in another. For our purposes we simply allow a user to designate a *different* set of data generation parameters for each region.

There are two major categories of data clusters as defined by the data generation parameters — *value clusters and density clusters*. A value cluster occurs when the differentiation of data characteristics is determined by *values* of fields of the data records defined in the region. For example, the values of the temperature field inside the cluster could be defined to have a mean of 34.5 with a standard deviation of 2.3, whereas outside the region, the mean might be 46.4 with a standard deviation of 5.6. A density cluster, on the other hand, is defined when the number of data records defined in the region has a significantly different density than the number of data records defined outside the region. For example, a cluster region could be defined by a range of temperatures between 0 and 32 degrees, such that the resulting data set should have approximately 3 data records per unit temperature range inside this region, but should average only 1 data record per unit temperature outside the region.

3.3 Formalization

Most scientific data can be described as unordered sets of multidimensional data. For the purpose of test data generation, we therefore assume a test data set to be an unordered set of n-dimensional data vectors (or data elements). Each data element can be seen as a point in n-dimensional space being defined along dimensions $x_1, x_2, ..., x_n$.

A cluster inside such test data sets can be defined as a set of points with some common characteristics that differ from the remaining points. A cluster may also be defined as a region in n-dimensional space with each of the data points inside the region having some characteristics that are clearly distinguishable from the rest of the data set. In this case, the cluster may be defined as a connected geometric object using a subset of the data dimensions. Sometimes, there may be no sharp border between the cluster region and the remaining data set. In this case, a threshold may be used to determine whether a data item belongs to the cluster or not. The dimensions that are used in the definition of a region are called *region dimensions*. If the region is defined by m dimensions, we call it an m-dimensional cluster where $0 \leq m \leq n$.

In addition to region dimensions, we also identify the dimensions that have the property of being *dense* such as the x and y coordinates in image data or the time dimension in time series data. We call such dimensions *array dimensions*. Without loss of generality, we assume that the first k data dimensions are the array dimensions $(x_1,...,x_k)$ and the dimensions $x_{k+1}, ..., x_n$ are the non-array dimensions. For each of the array dimensions (i=1..k), a range $[x_i^l; x_i^h]$ is defined with the number of data values in the range being n_i. Note that for each value $(v_1, ..., v_k)$ in the cross product of the ranges $[x_1^l; x_1^h] \times ... \times [x_k^l; x_k^h]$, there is exactly one data item in the data set that has $v_1, ..., v_k$ as the values for its first k dimensions. In other words, the first k dimensions are array dimensions if the projection of the n-dimensional data set onto the k array dimensions is bijective and the projection yields a k-dimensional rectangle covering each value inside that rectangle. In the case of using array dimensions, the number (N) of data items in the data set is given by the number of array dimensions and their ranges. It is the product of the n_i:

$$N = \prod_{n=1}^{k} n_i .$$

The array dimensions only contain information on the position of a data item inside the k-dimensional rectangle spanned by the ranges of the k array dimensions. By imposing an ordering on the data items and using the n_i as well as their ordering as meta-information, the same information is available without storing the array dimensions as part of the data vectors. For space efficiency reasons, many formats for storing data with array dimensions (e.g., image data) use some kind of convention which allows the array dimensions to be omitted.

In testing existing data sets for array dimensions, a necessary precondition that is easy to test is to get the ranges of each possible array dimension, to multiply the corresponding n_i, and to compare it with the number of data items in the data set. The sufficient condition for several dimensions to be array dimensions is much harder to test. It also requires a check for duplicate combinations of values in the possible array dimensions. In cases where no array dimensions can be identified, it may be interesting to extend or reduce the data set to allow some dimensions to be array dimensions. For this purpose, additional data items may be introduced using interpolation techniques or unnecessary and redundant data items may be omitted (or averaged). In some cases, it may even be desirable to turn data items with varying intervals between values into array dimensions. This can be done by artificially introducing an array dimension according to the ordering of the data items. The same can also be done for ordinal types whose data values are ordered but have no constant interval between values.

For visualization purposes, often a subset of the array dimensions is mapped to the dimensions of the visualization. Image data (#ArrayDimensions ≥ 2), for example, is usually mapped to the two dimensions of the display; time series of image data (#ArrayDimensions ≥ 3) are usually mapped to the two dimensions of the display plus time; time series of three-dimensional geometric data (#ArrayDimensions ≥ 4) are usually mapped to three display axes plus time, and so on. In these examples, the mappings are natural, but there are many other mappings possible, especially if $k \gg 4$ or $n \gg k$, which means that there are many more array dimensions than the three dimensions of the display plus time or that there are many non-array dimensions which are difficult to visualize if only the array dimensions are mapped to the three dimensions of the display plus time. For low array dimensionality ($k < 4$) or no array dimensions ($k = 0$), the task of visualizing the data is to find some meaningful mapping from non-array dimensions to the dimensions of the display plus time (which are basically all metric array-like dimensions in the visualization domain).

4. Test Data Generation

In generating multidimensional test data sets, it is important to distinguish data sets according to the number of array dimensions, the number of clusters, and the method used for describing them (data, value cluster or density cluster regions). All three aspects are important not only for determining the data generation parameters but also for the data generation process itself, especially for the constraints that apply in generating the data.

4.1 Constraints

Constraints in generating the data are especially important if one or more array dimensions are involved. One constraint is that the number of data items is given by the number and ranges of the array dimensions. Also, the number of data items for each data value in one array dimension is given as the product of the n_i of the remaining array dimensions. Similar constraints apply to any combination of the array dimensions. The constraints may also be expressed in terms of uniqueness and coverage of the value combinations for all array dimensions. The easiest way to fulfill these constraints is to generate the test data in an ordered fashion covering the allowed ranges for all array dimensions uniquely. An independent generation of the array dimensions would require checking the constraints for each generated data item which is computationally intensive. Still, in some cases it may be necessary to check some constraints. For example, if multiple region clusters are defined using array and non-array dimensions, then conflicts between the cluster definition and the constraints introduced by the array dimensions may occur.

4.2 Data generation parameters

Independently from the method used to describe the clusters, several data generation parameters are needed. Among the basic data generation parameters, there are the overall number of dimensions (n), the number of array dimensions (k) and their ranges, the number of clusters, and, in case $k = 0$, the number of data items. In order to generate test data, we need at least some more information about the non-array dimensions, namely their distribution function (uniform, normal, gaussian, ...) in case it is an independent dimension, or the correlation coefficient or functional dependency in case it is a dependent dimension. Array dimensions are considered independent dimensions which allows them to be used in defining the dependent ones. The different distribution functions are defined by specifying the necessary parameters: lower and upper limit for the uniform distribution, mean and standard deviation for the gaussian distribution, rho and lambda for the gamma distribution, and so on. Functional dependencies may be defined by an arbitrary function plus a randomness factor which is used to perturb the results of the functional dependency.

4.3 Cluster regions

A different way of describing the characteristics of the test data set is to explicitly define the cluster regions and their properties. Depending on the kind of clustering used, we distinguish between value cluster and density cluster regions. *Value cluster regions* are defined by identifying the region dimensions, defining the geometric shape of the region, the number or percentage of data items in the region, and the distribution function, correlation coefficient or functional dependency plus randomness factor for each region dimension. In our test data generation, regions are m-dimensional rectangles in n-dimensional space. This allows the regions to be defined by specifying some range for each region dimension. *Density cluster regions* are defined by identifying the region dimensions, defining the geometric shape of the region, and the density of elements in the region. The actual number of data items that are in each region and outside all regions is determined relative to each other.

The data items belonging to non-overlapping regions can be generated independently from each other. Regions that partially overlap with other regions require special consideration. The specified data characteristics for overlapping regions may be conflicting and may not be satisfiable by any data set. In order to interpret overlapping region specifications unambiguously, the order of defining the regions determines a priority ordering for the regions. The regions that are defined first have the highest priority. In case of cluster density regions, the regions which have the highest priority are filled with data items according to the desired density. For subsequent cluster density regions, only the non-overlapping part of the region is filled with data items according to the desired density.

We define the *base region* as the region in multidimensional space that includes all other regions. Assume, that the range of each region for dimension i is given by $[l_i, h_i]$. Then, the base region includes at least the multidimensional space defined by

$$[\min\{l_1\}, \max\{h_1\}] \quad x \quad \ldots \quad x \quad [\min\{l_n\}, \max\{h_n\}].$$

If some dimension is not used in any region definition, the range for that dimension is arbitrary. Note that in general, the base region is sparse since the number of data items may be low compared to its volume — it may even be empty.

Clusters that are defined using distribution functions such as normal or gaussian distributions provide smooth transitions into the region. Other cluster definitions, including density clusters may result in rather sharp transitions into the region. Such transitions may need to be smoothed to resemble real data. Defining smooth transitions into regions is not always straightforward, especially in the case of overlapping regions. We do not address this issue at the present time.

5. Examples

Tools that partially implement the described test generation facilities have been implemented at the University of Munich and the University of New Hampshire at Durham. The tool developed at the University of New Hampshire is described in a related paper in this volume [WB 94]. It is primarily oriented towards generating what we identified as *image data*. The tool developed at Munich focuses on the generation of *statistical data* as described in section 2. In the case of statistical data, the number of array dimensions is assumed to be zero. Different kinds of relationships may be defined between different dimensions in each of the clusters and the base region. Figure 1 shows visualizations produced by the VisDB system [KKS 93] using generated test data. The data of the four dimensions is generated such that only the first of the four dimensions is independent; the other three dimensions are functional dependant on dimension one. Dimension two is linear, dimension three quadratic, and dimension four is cubic dependant on dimension one. The generated data set consists of 6000 data items and the distribution of values for the independent dimension is uniform in the range [0, 100]. The data used to produce figure 1a has a randomness factor of zero which is increased to 0.5 in figure 1b and to 1.0 in figure 1c. Despite the linear functional dependency between dimension one and two, the corresponding visualizations in figure 1a are identical. This is due to the normalization and mapping of the different

a. Randomness Factor = 0.0   b. Randomness Factor = 0.5   c. Randomness Factor = 1.0

Figure 1: Visualizations from Test Data with Functional Dependencies

value ranges to a fixed color range. The main difference between dimension one and the dimensions with a higher order functional dependency is that the region of light coloring is larger. This is due to the unequal distribution of values in the extended value ranges of dimensions two and three. The increasing randomness factor results in some distortion of the visualization which also induces minor distortions in the visualization for dimension one. This results from a different ordering of data items which is caused by data items that have a high deviation from the functional dependency. More visualizations produced by the VisDB system using generated test data with different base region and cluster sizes can be found in a related paper of this volume [KK 94].

6. Conclusions and Future Work

In this paper we have described a model for test data generation that can be used to evaluate visualization techniques. The data sets are constructed from specifications that identify clusters of data that have different characteristics. Users can define clusters based on the density of data in the region or based on the values of the data. Statistical distributions, correlations, and functional dependencies can be used to determine the characteristics of the data in each region. Aspects of our model have been incorporated into two different systems for generating test data.

Our intent in defining our test data generation model is to begin to develop tools that can be used to provide support for rigorous evaluation of visualization techniques — especially those that present multivariate and/or multidimensional data. Our work is just a small step in this direction. The kinds of data sets that we can generate do not necessarily represent any particular kind of 'real data'. There are many other kinds of distributions that may be needed in order to provide truly meaningful tests for a particular domain. It would be nice provide arbitrarily shaped regions, to develop rigorous definitions of alternative interpretations of how to handle overlapping regions and to define smooth transitions across region boundaries. Finally, the most difficult work is the development a complete methodology for evaluating the effectiveness of visualization techniques.

Acknowledgments

This research has been supported in part by the National Science Foundation under grant IRI-9117153.

References

[Bed 90]   Beddow J.: *'Shape Coding of Multidimensional Data on a Microcomputer Display'*, Visualization '90, San Francisco, CA., 1990, pp. 238-246.

[BF 92]   Beshers C., Feiner S.: *'Automated Design of Virtual Worlds for Visualizing Multivariate Relations'*, Visualization '92, Boston, Mass., pp. 283-290.

[BF 93]   Beshers C., Feiner S.: *'AutoVisual: Rule-based Design of Interactive Multivariate Visualizations'*, Computer Graphics & Applications, Vol. 13, No. 4, 1993, pp. 41-49.

[Dev 87]   Devore J. L.: *'Probability and Statistics for Engineering and the Sciences'*, Brooks/Cole, Monterey, California, 1987.

[Dye 90]   Dyer D. S.: *'A Dataflow Kit for Visualization'*, Computer Graphics & Applications, Vol. 10, No. 4, 1990, pp. 60-69.

[GPW 89]   Grinstein G. G., Pickett R. M., Williams M. G.: *'EXVIS: An Exploratory Visualization Environment'*, Graphics Interface '89, London, Ontario, 1989.

[ID 90]   Inselberg A., Dimsdale B.: *'Parallel Coordinates: A Tool for Visualizing Multi-Dimensional Geometry'*, Visualization '90, San Francisco, CA., 1990, pp. 361-370.

[KKS 93]   Keim D. A., Kriegel H.-P., Seidl T.: *'Visual Feedback in Querying Large Databases'*, Visualization '93, San Jose, CA., 1993, pp. 158-165.

[KK 94]   Keim D. A., Kriegel H.-P.: *'Possibilities and Limits in Visualizing Large Amounts of Multidimensional Data'*, in: Perceptual Issues in Visualization, Springer, Berlin, 1994.

[LWW 90]   LeBlanc J., Ward M. O., Wittels N.: *'Exploring N-Dimensional Databases'*, Visualization '90, San Francisco, CA, 1990, pp. 230-237.

[Mih 90]   Mihalisin T., Timlin J., Schwegler J.: *'Visualizing Multivariate Functions, Data and Distributions'*, Computer Graphics & Applications, Vol. 11, No. 3, 1991, pp. 28-35.

[PG 88]   Pickett R. M., Grinstein G. G.: *'Iconographic Displays for Visualizing Multidimensional Data'*, Proc. IEEE Conf. on Systems, Man and Cybernetics, Beijing and Shenyang, China, 1988.

[SGB 91]   Smith S., Grinstein G. G., Bergeron R. D.: *'Interactive Data Exploration with a Supercomputer'*, Visualization '91, San Diego, CA, 1991, pp. 248-254.

[SP 82]   Swets J. A., Pickett R. M.: *'Evaluation of Diagnostic Systems: Methods from Signal Detection Theory'*, Academic Press, New York, 1982.

[Ups 89]   Upson C., et al.: *'The Application Visualization System: A Computational Environment for Scientific Visualization'*, Computer Graphics & Applications, Vol. 9, No. 4, 1989, pp. 30-42.

[WB 94]   Wong P. C., Bergeron R. D.: *'A Multidimensional Multivariate Image Evaluation Tool'*, in: Perceptual Issues in Visualization, Springer, Berlin, 1994.

Interaction in Perceptually-Based Visualization

W. Hibbard[1], H. Levkowitz[2], J. Haswell[3], P. Rheingans[4], and F. Schroeder[5]

[1] University of Wisconsin Madison
[2] University of Massachusetts Lowell
[3] Rutherford Appelton Laboratories
[4] Martin Marietta US EPA Visualization Center
[5] Fraunhoffer Computer Graphics

1 Introduction

The purpose of visualization is to present information to human beings. Perceptually-based visualization aims at making such presentation as efficient and effective as possible, from the human's point of view. I.e., it seeks to maximize the amount of information that the human viewer can perceive out of the presentation.

To achieve that, we need to understand perception, and apply it to all visualizations.

Perception is an active process: it is more than the passive accepting and processing of the raw input from our senses by a genetically determined hierarchy of standard engineering analyses (filtering, summation and averaging) embodied in the nervous system. There are some "built-in" characteristics of human sensory systems: for example, the range of wavelengths we can see and hear, the colors or tones that we can distinguish, the perception of apparent smooth movement that we get when still pictures are sequenced rapidly enough. But there is more to perception than this; our brains interactively explore raw sensory inputs in order to fit them into a mental model of the world. For example, our two eyes see slightly different views of the same scene, and our brains create links between corresponding points in these two views by a trial and error process. In general, our higher mental functions do not assimilate all of the raw sensory information at once. Rather, information flows in both directions between higher mental functions and the sensory input, creating a feedback loop. It is natural to view interactive techniques as an extension of this feedback loop outside of the brain, where interactive computer systems invoke perceptions in response to users' controls. Such a model-based approach to perception is supported as a valid approach by the success of parallel work in artificial intelligence by Marr and his co-workers [9]. These machine-based models open up the possibility of testing out the relative plausibility of competing theoretical models of perception.

Interaction is indispensable for perception. At the individual level, our vision depends on interaction; we change our viewpoint to help resolve the ambiguity in the 2D views our eyes have of the 3D world. We do this by moving our heads to create motion parallax to better understand the 3D shape of objects around us. Motion parallax uses the difference in the apparent direction of an object as seen from several points that are not colinear with the object to enhance the three-dimensional shape perception of the object.

At the highest level, humans are engaged in understanding the world. However, there is more raw data about the world than people can directly perceive. Scientific experiments collect and analyze some particular raw data, in order to build models that apply in general ways to the world. These models are tested by their ability to predict other phenomena. This is really a form of interactive perception, where changes in the way data are collected and analyzed affect the way they are perceived. This is true for any information that is presented to a human being; one needs to interact with the data to bring it to a form where one can analyze and perceive it.

While interaction is essential for perception, it is also necessary for the development of perception in childern and young animals. Experiments with children, animals and adult humans have shown that, apart from some simple processes, such as figure-ground distinction, higher animals need to interact with their environment to learn how to process visual input. This idea has been present in the theories of visual perception advanced some years ago by, for example, Gibson [5]. Informal support is offered by anthropological studies of (pictorial) representations used by different groups and the difficulty of understanding novel representations. This is supported by experimental studies of environmental effects upon the strength of standard visual illusions in groups with different environmental surroundings. Visual deprivation studies, such as those physiological experiments carried out by Blakemore and his co-workers on cats (following on from earlier "kitten-carousel" studies), strongly suggest that there is some form of tuning-in of the mammalian visual system early in the development [1]. In these experiments, two groups of kittens saw the same visual stimulus. However, one group was restricted in its ability to interact with, and thus control the visula input. As a result, that group's perception development was inferior to that of the group whose interaction was not restricted.

In this paper, we summarize the main issues that involve interaction with data and perception of information in the data. Section 2 describes the perceptual properties of interaction. Section 3 discusses modes of interaction. Section 4 presents a taxonomy of the loci of interaction, i.e. the location where users perceive excercising control over the interaction. We conclude in Section 5.

To obtain an in-depth understanding of general human factors issues refer to [2, 3, 4] while [6, 7, 8, 10, 13] concentrate on visual perception. For more information on direct manipulation and response rates see [14].

2 Perceptual properties of interaction

Just as the understanding of human perception provides guidance for the design of systems that produce images and sounds, it can also provide guidance for the design of interaction techniques in such systems.

2.1 Characteristic times for perceptual tasks

Although the speed of computers is constantly increasing, the speed of our brains is either fixed or evolving very slowly. Thus we can identify characteristic times for perceptual tasks. These characteristic times define response time requirements for the design of interactive computer systems. We should note that these times vary among people and depend on the details of the perceptual tasks.

Like movies and television, computers generate moving images as sequences of still frames. If the still images are presented at too slow a rate, we perceive them as a sequence of images rather than as continuous motion. The threshold for apparent motion, i.e., the rate at which a sequence of still frames creates the perception of continuous motion is 16 frames per second; however, at this rate, motion still appears a little fragmented. The motion picture industry uses the rate of 24 frames per second, which provides the perception of fully continuous motion. However, the apparent motion threshold is only one factor; as a sequence of frames is presented, the screen is blanked every time a new frame is introduced. Thus, the brightness of the screen fluctuates between bright and dark, creating flicker. The flicker fusion rate, which is the rate above which no flicker is perceived by the viewer, is around 60 frames per second, corresponding to about 1 frame per second. Human sensitivity to flicker peaks at about 10 frames per second. To avoid flicker, the motion picture industry uses shutters to double or triple the frequecy of bright/dark changes. In CRT displays, 30 frames per second displays use interlacing to double the flicker frequency. Alternatively, non-flicker displays exist with rates of at least 60 frames per second.

A system that produces images "on the fly" and in response to user controls must produce a new frame every one thirtieth of a second in order to create the perception of motion. It may be possible to pipeline the production of frames somewhat, but a pipelined output of images is not really interactive. In such a case, the system's response to a user's control will not appear in the frame following the one in which the control is exercised. Response time requirements for controls such as 3D rotations are not as well defined as those to achieve apparent motion and to overcome flicker.

In addition to the need to provide flicker-free, continuous apparent motion, head-mounted virtual reality displays must produce images at rates fast enough that users will not perceive a lag between head rotation and scene rotation. This rate is about 60 frames per second—the same rate required for no flicker, corresponding to about 0.016 seconds per frame. This response time is required from a user control to obtain the desired perceived effect, which precludes the pipelined production of frames. Thus, the nature of human perception places much higher requirements on head-mounted virtual reality systems than on interactive visualization or animation systems.

Other characteristic times of human perception are more subjective, but no less important for successful systems. For example, in the task of text editing we can measure the time for text to appear or disappear, or for the cursor to move, in response to user keystrokes. Tolerance levels vary between people, but our efficiencies at text editing degrade significantly when our tolerances are exceeded. A response time tolerance of one half a second is typical. This is about the time it takes for a non-touch typist to look from the keyboard to the screen to verify that one task has been accomplished correctly (e.g., the cursor is in the right place) before beginning the next task (e.g., inserting a word).

Users require more "think time" for higher level tasks, and will tolerate similarly longer periods of time for system responses. We can define the characteristic time for a high-level task as the duration it takes for a user's mind to wander off the task while waiting for the computer's response. Beyond this time users will shift their attention to other tasks (e.g., they may start reading their email). People can accomplish tasks without continuous attention, but switching attention radically changes the efficiency of problem solving.

For many perceptual tasks, consistency of response times is as important as fast response times. That is, users are more willing to adapt to slow systems than they are to adapt to systems that are sometimes fast and sometimes slow. This suggests that designers may sometimes improve systems with a high variance of response times by delaying fast responses to conform to slower responses.

2.2 The effect of response time on understanding controls

In Section 2.1 we described some immediate effects of system response time on human perception. Here we describe a secondary effect of system response time. An interactive system presents users with a variety of controls, and it is common for users to not understand the effects of all the controls. One important way for users to learn the effect of controls is by observing how the system responds to them. The perceived link between a control and its effect will be strongest if the response time is short. During a long response time, a user may either switch attention and forget which control was used, or may exercise another control and thus lose the one-to-one correspondence between controls and their effects.

Since users in many applications are constantly being confronted with new interactive systems; the general issue of how to make their controls easier to understand is an important research area.

2.3 The tradeoff between speed, aesthetic quality, and fidelity

There is a tradeoff between the number of operations required to compute a presentation (e.g., an image or sound), the aesthetic quality of the presentation, and the fidelity of the presentation to the object it represents (e.g., to the data being visualized). In Section 2.1 we discussed several limits on the time available for computing certain presentations. For a given computer system, the number of operations is proportional to computing time, so these time limits imply limits on the quality and fidelity of presentations. The fidelity of a presentation is the accuracy with which it presents the contents it is presenting. By aesthetic quality, we mean the rendering quality. For example, if a system can render 60,000 polygons per second and we want to use it for generating a head-mounted virtual reality display, then we must produce 60 frames per second for each eye. Thus, each frame will be able to contain at most 500 polygons. This can potentially limit both the quality and the fidelity of the display.

A variety of techniques can be used to provide reasonable response rates with minimal reduction in quality or fidelity. One technique is progressive refinement: images are first rendered at a low detail level, and then progressively rendered at higher levels of detail. When images are being animated using progressive refinement, only the low detail versions appear, but when animation stops, the low detail images are replaced with higher detail versions. A second technique is to use low detail images to allow the user to plan a "flip book" animation sequence of high detail images (a "flip book" animation is created from a sequence of frames pre computed and stored in the computer, rather than by computing the frames in real time as they are displayed).

2.4   The effort required for exercising controls

As we have discussed, the time it takes for a system to respond to controls affects our perception of those controls. However, we also have innate perceptions of controls. One of these relates to the amount of effort required to exercise a control. Controls that are difficult to exercise may cause users to switch attention from their task to the details of the control, or to avoid using them altogether. For example, if a control requires a double click on a mouse at a speed significantly faster than a user normally double clicks, the user will need to switch attention away from the current task to make a fast double click. Controls may be difficult to exercise because a single operation at the user's task level may require a long sequence of controls at the system level. For example, selecting an operation may require the user to descend through a long menu hierarchy, or rotating a scene in three dimensions may require the user to compose a sequence of 2D rotations. Controls may also be difficult because they require the user to switch between multiple input devices (e.g., between mouse and keyboard).

The point of this discussion is not to condemn all complex controls—sometimes it is hard for system designers to avoid them. Rather, the point is to recognize that there are perceptual effects of the difficulty of exercising controls, which can interfere with users' attention to their tasks.

One interesting example of difficult controls is provided by an old system for plotting points on maps. Points were specified using a cross hair on a map table, and sampled by pressing a button on a console that was out of reach of the map table. The result was that the system required two operators and verbal communications between them in order to plot points.

2.5   The breadth of choice for controls

When users have complex tasks, they often have to make choices from very broad sets of operations. This can be a valid motive for controls that are difficult to exercise. However, the breadth of choice of controls has an innate perceptual effect, simply because we have to think harder to choose from larger sets.

The recent emphasis on integrating visualization with data analysis provides a good example of very broad choices of controls. The choice of data analysis functions is extremely broad—it is as broad as the choice of computer algorithms. Thus, the problem of controlling analysis functions is coming to dominate the problem of controlling visualizations. This is particularly true in systems where the same controls are used for both visualization and analysis. For example, data flow networks in several systems mix analysis and visualization modules, and visualizations are designed as networks. The breadth of user choices has become an important research problem for designers of interactive visualization systems.

2.6   The naturalness of controls

Controls are physical actions (e.g., pushing on keys or mouse buttons, moving a mouse, moving a joystick, speaking, etc.) that represent conceptual actions. Some physical actions are more intuitive than others for representing given conceptual actions. This is a very complex issue, and a rich source of research problems in perception. Some controls seem to be easier to exercise because we perceive the physical action as analogous to the conceptual action it represents. For example, many people (but not all) prefer to control 3D rotations by dragging a mouse over a 3D scene in a direction parallel to the desired 3D rotation—the physical action is parallel to the conceptual action it controls. Moving a file icon between two directory icons by dragging it with a mouse provides a similar parallel between physical and conceptual actions.

The user's acceptance of the mouse for 3D manipulations may be related to the fact that he or she may already use it for many 2D interactions. Most visualization tasks require a mixture of 2D and 3D operations; switching between devices should be kept to a minimum to avoid confusion. However, although 2D tasks can be simulated with 3D devices (such as the Spaceball), many users might be reluctant to change over to a completely new device. This is similar to experienced typists, who often prefer keyboard commands to point-and-click mouse operations.

It is common for the realities of computer systems to introduce constraints in the sequence of controls that are not analogous to any conceptual constraint. For example, a system that is capable of interactive 3D rotations may also include a function for producing animations as flip books of rendered images. To address the speed-versus-quality tradeoff discussed in Section 2.3, the user might be able to rotate a still 3D scene, or animate it, but might not be able to rotate an animated 3D scene. This constraint on controls has no conceptual analog that would preclude animated scenes from being rotated. The reason for such a constraint may be purely impelemntational, or it may reflect the assumption that users may encounter difficulties following simultaneous changes in so many variables of the display. On a typical 2D, flat screen the brain attempts to build up a 3D mental image of an object from the rotations. However, if the object is also changing it may be difficult to resolve the two effects. Sufficient time needs to be given to comprehend the changes in the objects in a scene, before moving to a new viewpoint. If the changes follow a cyclic or obvious pattern it may then be possible to rotate while the animation continues since the viewer no longer needs to concentrate on the changing objects.

2.7 The vocabulary of controls

This issue relates to the naturalness of controls, focusing on how well the vocabulary of controls corresponds to the vocabulary of the user's tasks. If every control corresponds to some basic operation of the application field, then the application area gives the user a conceptual framework for understanding the controls.

3 The modes of interaction

In Section 2 we described some basic perceptual properties of interaction. In this and the next section, we define taxonomies of interaction techniques. We assume that a computer is transforming an object of perception into some depiction sensed by the user. The object of perception may be data, a process (such as a running algorithm or a running nuclear power plant), or a real object being sensed by instruments under computer control. The depiction may be images, sounds, feelings, tastes and/or smells. It may be static or dynamic (e.g., an animated sequence of images). Given these terms, we can define a simple taxonomy of basic modes of interaction.

3.1 Controlling the object of perception

In this mode of interaction, the user changes the object of perception. Inserting text in a text editor is one example of this mode of interaction. The depiction changes in response to the control because the object being transformed into the depiction changes. Since the object of perception may take such a variety of forms and values, we do not classify this mode of interaction further.

3.2 Controlling the transformation of an object into a depiction

In this mode of interaction, the user controls the way the object of perception is transformed into a depiction but does not change the object of perception.

Controlling the functional form of the transformation. We assume that the transformation from an object of perception to a depiction is expressed as a complex composition of basic functions (i.e., not necessarily a linear composition of functions). In this mode of interaction, the user changes the format of the composition of functions that defines the transformation. In some visualization systems, the transformation of data into images is expressed as a data flow diagram of basic modules. Editing the data flow diagram is an example of controlling the functional form of the transformation of data into images.

Controlling parameters of the transformation. This mode changes parameters of the transformation of the object of perception to depiction, but does not change the functional form of the transformation. Changing an iso-level of a 3D field depicted by an iso-surface is an example of this mode of interaction. Changing color maps used in the production of images is another example.

Of course, a change of functional form can often be expressed as a mere change of parameters of a more complex functional form. However, in practice the distinction between changing parameters and changing functional forms is usually clear.

Some researchers are investigating the use of the tactile sense for understanding data. For example, a user may move his or her hands and fingers over a depiction of an object. In this case, the depiction is the pattern of forces applied to the user's hands and fingers. This is a special case of changing the parameters of transformation.

Controlling the projection from 3D to 2D in vision. Controlling the projection from 3D to 2D is a form of controlling the parameters of transformation, but it is such an important mode that we discuss it separately. Humans (and other animals) sense a 3D world using 2D arrays of sensors in their eyes, so raw visual perceptions are necessarily ambiguous. Our brain constructs a 3D model of the world from 2D views. These mental processes are fundamental to perception. Thus, we classify the control of the 3D-to-2D projection as a special mode of interaction. For example, when projecting a cube using a wire-frame drawing, the correct perception of the cube may depenend on the projection; projecting down the main diagonal of the cube will produce a hexagonal disc, which may be perceived both as a 2D object and as the projection of a 3D object. However, a slight change in the direction of projection will reveal the nature of the object.

Controlling the selection of subsets. One of the most common interaction techniques is selection of subsets. Most often, we select spatial subsets at different resolutions but at the same location when we zoom, whereas pan changes the location of the subset but preserves the resolution. These modes of interaction are also special cases of controlling the parameters of transformation, since they are just linear transformations of our raw perceptions. However, they are often coupled with "intelligent" versions, which change the functional form of the transformation. Fine detail in the object of perception may not have the same form as coarser information. When fine detail crosses the threshold of perceptibility during a zoom, a new functional form of transformation may be necessary.

These spatial subset selection techniques can be extended to non spatial (and other) domains. For example, it is possible to "zoom" or "pan" in color space; a color space zoom introduces more (or fewer) distinct colors within a range of values along a color axis. A pan in color space selects different sets of colors, each with the same number of distinct colors. These concepts are also being extended to other domains, such as sound [11, 12].

4 The loci of interaction

In this section we present a taxonomy based on the location where users perceive that they exercise control.

4.1 Control located in depiction

The user can perceive that control is exercised at a location in the depiction of the object of perception. The user has the impression of interacting directly with the depiction, particularly when the physical action of control is analogous to the conceptual action, as discussed in Section 2.6. An interesting example of locating the control at the depiction is the tactile depictions described in Section 3.2.2.

4.2 Control located in icons separate from depiction

The user may perceive control as exercised at a location separate from the depiction, in some kind of a control icon. Control icons may exist on a display screen or in a virtual graphical space, may be physical devices (such as joysticks or spaceballs), or may be voice activated. Rotating or resizing an object by specifying the transformation via sliders is a control that is excercised at a separate location. This is a significantly different interaction from an interaction that uses direct jestures on the object.

5 Conclusions

Perception is active even inside our brains, and interaction with the outside world is essential for effective perception. Thus, interaction must be part of our study of perception. The nature of human perception has numerous implications for the design of interaction techniques. There are characteristic times for certain perceptual tasks that determine response time requirements for interactive systems. Perceptual issues must also be considered in the design of easy-to-use controls for interactive systems. Thus, just as an understanding of human perception can help us design images and sounds that depict data effectively, this understanding can also help us design better interactive techniques for visualization and sonification systems.

We have summarized here the various aspects of the influence of interaction on perception, and thus on effective perceptually-based presentations of information. Such interactions with the data will become more and more essential as users interact with larger, more complex data sets, in particular during exploratory visualizations, i.e., visualizations where the user has no previous knowledge of the contents in the data.

Acknowledgments

Thanks to A. Conway of Rutherford Appleton Laboratory for providing the information about the kitten experiments.

References

1. C. Blakemore and G.F. Cooper. Development of the brain depends on the visual environment. *Nature*, 228:477–478, 1970.
2. K.R. Boff, L. Kaufman, and J.P. Thomas (eds.). *Handbook of Perception and Human Performance. Vol I—Sensory Perception and Human Perception.* John Wiley and Sons, Inc., New York, 1986.
3. K.R. Boff, L. Kaufman, and J.P. Thomas (eds.). *Handbook of Perception and Human Performance. Vol II—Cognitive Processes and Performance.* John Wiley and Sons, Inc., New York, 1986.
4. M. Helander (ed.). *Handbook of Human-Computer Interaction.* Elsevier Science Publishing Company, Inc, 1988.
5. J. J. Gibson. *The Perception of the Visual World.* Houghton-Mifflin, Boston, MA, 1950.
6. R.L. Gregory. *The Intelligent Eye.* Weidenfeld and Nicolson, London, England, 1970.
7. R.L. Gregory. *Concepts and Mechanisms of Perception.* Gerald Duckworth and Co. Ltd., London, England, 1974.
8. E.C. Hildreth. *The Measurement of Visual Motion.* MIT press, Cambridge, MA, and London, England, 1984.
9. D. Marr. *Vision.* W. H. Freeman and Company, New York, NY, 1982. ISBN 0-7167-1567-8.
10. S. Pinker (ed.). *Visual Cognition.* MIT press, Cambridge, MA, and London, England, 1985.
11. K. Seetharaman. *An interaction model for exploratory data visualization.* PhD thesis, University of Massachusetts Lowell, August 1994.
12. K. Seetharaman, G. Grinstein, H. Levkowitz, and R. D. Bergeron. A conceptual model for interaction in multiple representational spaces. In *International Conference on Computer Graphics: Interaction, Design, Modeling and Visualization,* Bombay, India, Feb 22–26 1993. International Federation for Information Processing and Computer Society of India. Submitted.
13. R.N. Shepard and L.A. Cooper. *Mental Images and Their Transformations.* MIT press, Cambridge, MA, and London, England, 1982.
14. B. Shneiderman. *Designing the user interface.* Addison-Wesley, Reading, MA, 1987.

Harnessing Preattentive Perceptual Processes in Visualization

Ronald M. Pickett[1], Georges Grinstein[1,2], Haim Levkowitz[1], and Stuart Smith[1]

[1] Institute for Visualization and Perception Research, Department of Computer Science, University of Massachusetts Lowell, Lowell, MA 01854, USA
[2] The MITRE Corporation, Center for Airforce C3 Systems, 202 Burlington Road, Bedford, MA 01730-1420

Abstract. We present an overview of our visualization work at the University of Massachusetts Lowell. We first describe our general approach of creating iconographic displays and the perceptual rationale for how we design those displays. We then describe our accomplishments to date, mainly to develop a general purpose system for creating iconographic displays and to produce illustrative displays with different types of icons and on a wide variety of databases. We conclude with a description of two main long-term goals: (1) to develop a more capable display system; and (2) to conduct basic applied research.

1 Introduction

Our work in visualization focuses on the development of displays and interaction techniques that will enable scientists to explore large multidimensional databases by visual and other perceptual means. We seek to facilitate the exploration of various types of empirical data, including large statistical databases such as those obtained for studies of health and crime statistics, as well as multiparameter image data like those obtained by earth resources satellites and medical imaging systems. The current work in visualization at the University of Massachusetts Lowell has been underway for a little over five years [1–26]; however, much of the general rationale and technical approach can be traced to earlier work elsewhere [27–33]. In this paper, we first provide background on the technical approach and accomplishments of our work to date. We then briefly summarize our long-term goals.

2 Perceptual Rationale

Though visualization technology has improved enormously over the last decade, the improvements have been mostly enhancements of familiar and conventional forms of display. What has been lacking, and what, we believe, will preclude further large strides in visualization, is a quest for new forms of display. New forms of display will come from first thinking about the kinds of perception potentially to be harnessed in the service of data exploration, then from considering the kinds of displays needed to evoke those perceptions.

2.1 Harnessing Low-Level Perceptual Processes

Our intent is to create displays that make structures in data perceptually apparent. We are particularly concerned with harnessing low-level perceptual processes, including aspects of texture, color, motion, and depth perception. The processes of interest are those that equip humans for automatic, real-time responses to objects and events visually encountered in natural ecological settings. Central to our rationale is the need to create "data pictures" with such physical verisimilitude that they trick the perceptual processes into an automatic attempt to make physical sense of them.

2.2 Iconographic Display Techniques

Our general approach to creating these perceptually compelling displays has been to adopt an iconographic technique. In this technique, each data item is encoded into a compact graphic element, or icon, encompassing an area of pixels. Many kinds of features of the icon can be varied under data control. Large samples of data are presented by displaying the icons en masse in a densely packed display. Which visible features of an icon are put under data control and how the icons are massed depend on the kind of low-level perceptual processes one seeks to evoke. Our approach to date has been to amass the icons into surface texture displays. When a large number of small visible elements are densely packed onto a surface. the resulting visual impression is that of a surface texture. If, for contiguous subsets of the elements, certain features of the elements differ significantly from those in other contiguous subsets, the visual system will automatically segment the display into regions of different texture. When the texture elements are icons into which data have been coded, these segmentations provide a basis for seeing statistical structure in the data.

Stick-Figure Icon Displays. For the texture elements, we have used primarily a small stick-figure icon, which consists of connected straight-line segments—a "body" segment and as many as four "limbs" (see Figure 1). The values of each data item are mapped to the segments and control any one of four visible features of the segments—orientation, length, brightness, and color. Because variations in line segment orientation have been shown to be a particularly potent basis for surface texture segmentation, we have worked with segment orientation almost exclusively. The possible patterns of connection of the limbs to the body or to each other—in sequence or in parallel—define the 12-member family of variations shown in the figure, any one of which can be chosen to represent a given set of data.

Demonstration with Multiparameter Imagery Data. A difficult problem in the analysis of multiparameter images is how to display them in such a way that the spatial patterns of relationship among the parameters are amenable to visual analysis. Showing the gray-scale images of the parameters separately in a side-by-side display provides for only the very crudest of visual analyses. What is needed

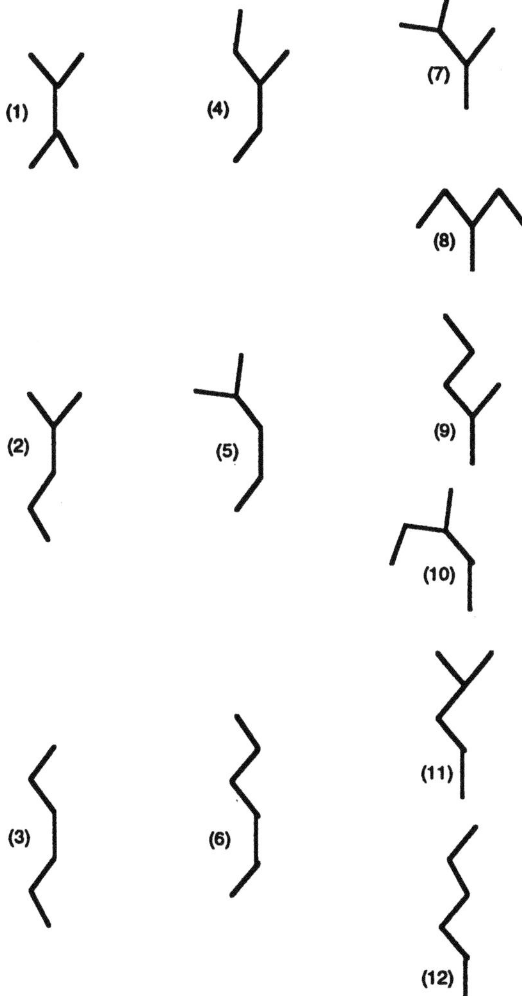

Fig. 1. Stick-figure icon family. Each member has four limbs, the orientation of which are under data control. The different patterns of attachment of the limbs make each member potentially suitable for conveying different types of data structure.

is a display that integrates the information from multiple images into a single picture. This can be done very effectively with color coding, as we explain further below, but only for up to three parameters. Figure 2 shows how integration of five parameters can be achieved with the stick-figure icon. This picture of multiparameter satellite imagery well illustrates the potential for segmenting an image very finely on the basis of texture differences and clearly suggests that with the stick-figure icon we can provide effective integrations of more than three parameters.

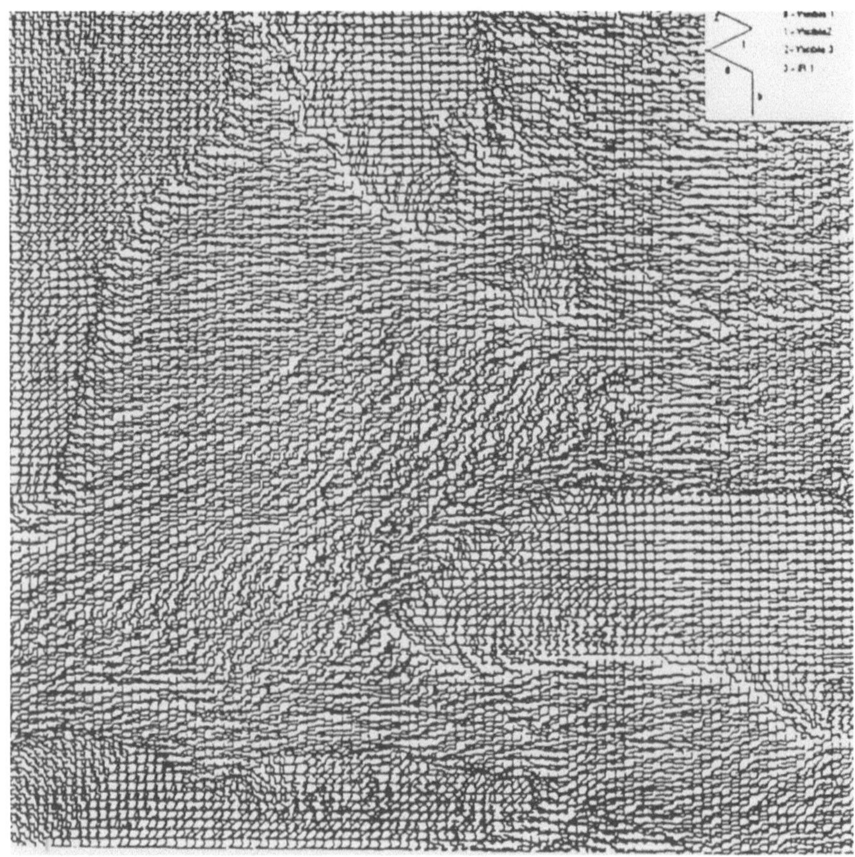

Fig. 2. An iconographic image of five-parameter weather satellite imagery of the Great Lakes region. Icon No. 12 was used. and the insert shows the particular mapping of parameters to icon segments.

Other Icon Forms. Icons can have many different kinds of data-controllable features. As already noted. the stick-figure icon can provide for data control not just over the geometrical aspects of the segments—their orientation and length—but also over their brightness and color. Color is a particularly powerful feature by itself. but can also be effectively integrated with geometric features. We discuss these possibilities first and then proceed to suggest yet other possible icon forms and features.

Color. Color, as noted above. is a very effective way to merge multiple parameters of imagery data. For up to three parameter images, the values of the pixels in the separate images control the three color coordinates of the corresponding pixel in the integrated display. Different combinations of values and different color models provide a wide range of options.

Sound. Data can also be encoded as the values of sound attributes such as pitch, loudness, attack rate, and decay rate [6,11,16]. Scientists can then listen to streams of sequential data, such as an EKG record, or a stock market history. Sound can also be added as an attribute of visualized data. Users can trigger data-controllable auditory features by "plucking" or "strumming" the icons with the mouse cursor, or these features can be initiated by some automatic process. When integrated with the visual representation, sound representation can be used to reinforce the visual representation or to encode additional data parameters.

2.3   Perceptions of Physical Characteristics

Very important to our perceptual rationale are techniques for making the icons appear more physically real. We are very interested, for example, in creating a three-dimensional version of the stick-figure icon so that segment orientation can be controlled in 3-D and presented, for a yet more realistic impression, via a stereo display. We also seek to create icons with surfaces or facets, in which the translucence and reflectance of the facets, as well as their shape and orientation, are data-controlled. These faceted icons can then be rendered realistically with light modeling techniques. We are also interested in the possibility of creating icons that have visually conveyable physical properties. We envision, for example, a cantilever beam icon in which physical parameters that affect how it bends and twists under different conditions of loading, such as its modulus of elasticity, cross-sectional area, and cross-sectional shape, are under data control. These beams can provide a natural, physical form of display, appearing en masse as a kind of grassy texture. When viewed from different angles or when viewed as the direction or degree of gravitational force is varied (see section 2.3), they could potentially provide a very compelling sense of high order structure in the data.We are not now capable of producing such a fully realistic cantilever beam display, but we have created an approximation to it with our "velcro" icon. This is a line icon implanted at one end in a planar surface. With its x, y location in the plane, it implant angles to the plane as well as its length and curvature under data control, this icon can encode up to eight data dimensions. When rendered in 3D and in stereo (see Figure 3), it provides a good impression of what our hypothetical cantilever beam display would look like. The data of Figure 3 are the same five-parameter satellite imagery data as in Figure 2.

2.4   Other Forms of Low-Level Perception to be Harnessed

We can create displays in which the icons move about under data control. We can imagine, for example, icons that "walk," "swim," or "fly" under data control.

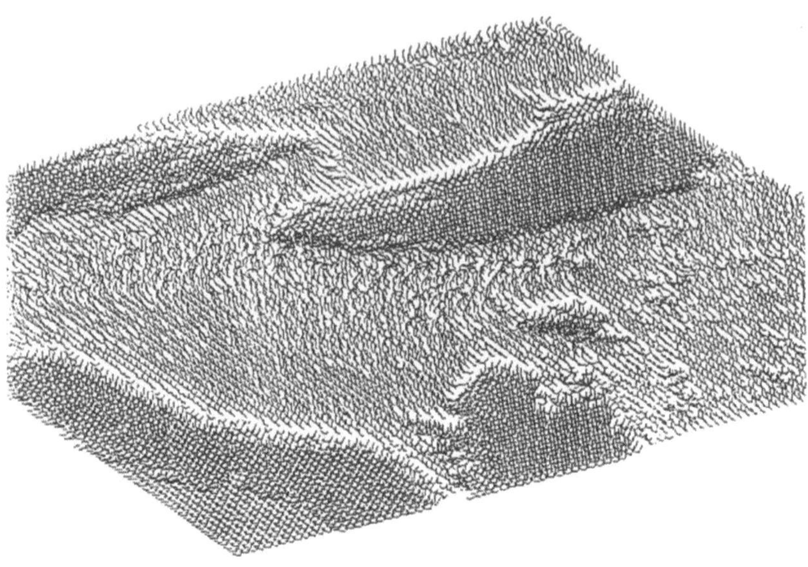

Fig. 3. Iconographic display of the same satellite imagery data of Figure 2. Here we employ the "velcro" icon.

When displayed en masse, such icons could evoke impressions of such visually interesting qualities as flowing, churning and swarming. While very little is known about these kinds of perception, they merit serious consideration because of their potential power for revealing subtle and complex structures in data.

2.5 Physical Interactions with Iconographic Displays

By interaction, one usually means interaction of the user with the system. This conventional kind of interaction is very important to consider. and we have been deeply concerned with it, both in creating the present system and in thinking about future systems. Much more critical to the present discussion, however, is interaction of a very different and unusual kind: interaction between the user and the data, or more precisely, between the user and the apparent physical objects, substances, and processes that represent the data. We will concern ourselves here exclusively with this latter kind of interaction. A central aim of our work is to make data look like real objects and materials that the user can reach out and touch. What we consider now is the possibility of actually allowing the user to do that in various ways. This touching would, of course, have to be electronically mediated, but it could still be very convincing. This kind of "physical" interaction could make the display more effective in two ways. First, it could provide additional channels for presenting data. We have already mentioned the capability to elicit auditory attributes by plucking or strumming the icon with the cursor. But, let us consider other possibilities. For example, we can imagine providing the cantilever beam icon with tactile properties. When the cursor hits such an icon, tactile feedback could be provided. To bend stiffer beams one would have to push harder to move the cursor. The iconographic display could thus create for the user a coordinated intersensory experience in the realms of vision, sound, and touch. The second way in which physical interaction could increase the effectiveness of the display is by making the display more perceptually compelling. We can imagine a surface texture display of densely packed cantilever beams sitting on a platform that the user can move. As the user tilts the platform, the beams tilt in response to the change in direction of the gravitational force, giving the user an immensely richer visual impression of the data-controlled mechanical properties of the beams. We can also imagine similar interactions with a display of icons with illuminated facets. The user could move the display much as one might rotate and tilt a piece of sparkling ore in sunlight. The surface would sparkle in informative dynamic reactions to the motions. We think that such displays with such interactions could provide highly effective new media for exploring data.

3 Accomplishments to Date

3.1 Display System Development

We have developed a prototype visualization system, *Exvis*, that allows users to display iconographic pictures of data sets using the stick-figure icon, the "velcro" icon, color model coding [14, 24], and the color icon [15]. In addition, the system allows coding of data into sound [17, 27]. This system provides capabilities for controlling such display attributes as the size of the icon bounding box, the size of the icon limb, the base orientation of the icons, the limits of angular travel of the icon limbs, the maximum amount of random displacement—or *jitter*—of each icon in the x and y directions. the color model(s) used, and the methods for blending colors within icons. These controls allow us to fine-tune iconographic pictures and obtain displays that show parameter interactions more clearly. Because users benefit greatly from the ability to observe changes in data pictures while they are manipulating these controls, we have implemented a subset of Exvis on a Thinking Machines Corporation CM-2 Connection Machine, which provides the computational power needed to support this kind of real-time interaction [17]. To make Exvis more accessible, we have also ported a subset to the Macintosh environment.

3.2 Demonstrations with a Wide Variety of Databases

We have felt it important as we went along to become familiar with as wide a variety as possible of data types and user needs. To that end, we have sought to acquire databases and work with scientists in a wide variety of settings both within and outside the University. Table 1 shows a list of databases with which we have worked. In each case, we have created the display and then have interacted with the scientists involved and with the displays to find out what they show and how we might improve them. We are currently acquiring other databases, including seismic data from two petroleum companies and, from our School of Management Science, a large statistical database on failing businesses.

3.3 Videos

We have produced three video cassette demonstrations of our work at different stages [12, 17, 18].

4 Long-term Goals

We have two main long-term goals. The first goal is to develop a more capable system for producing the displays and for providing interesting, perceptually stimulating interactions. The second goal is to develop a better understanding of these new display and interaction techniques—how they work, how well they work, and how to improve them—through a well-guided program of research.

4.1 Develop a More Capable System for Display and Interaction

To carry out the work described here, we need a system much more capable than the one we presently have. Our requirements include both hardware and software. The principal hardware requirement is a color graphics system capable of supporting real-time interaction with the geometric and color displays we have already developed, as well as with dynamic, intersensory displays such as the cantilever beam and illuminated facet displays described above. Among the software items we must develop in the near term are:

- Modular Version of Exvis. Exvis must be redesigned as a set of interactive modules, each of which implements just one data encoding technique (stick-figure icon, color icon, cantilever beam icon, illuminated facet icon, etc.). The modules would have a consistent user interface so that a user familiar with one module can move easily to any of the others.

Multiparameter Imagery

Subject

Subject	Source
Weather Satellite Images (2 IR and 3 Visual Bands)	US Air Force Phillips Laboratory
Medical Images—Abdomen (CT, MR_1, MR_2)	Harvard Medical (BWH)
	University of Massachusetts Medical Center
Medical Images—Brain Stem (4 parameter MR)	Harvard Medical (MGH)
Military Surveillance Images (Visual and IR)	Litton/Itek
Semiconductor Wafer Tests (Various Electrical Response Parameters)	Digital Equipment Corporation
Plasma Flow Images (Various PDE Parameters)	University of Massachusetts Lowell Department of Mechanical Engineering

Statistical Databases

Subject

Subject	Source
Epidemiological Data on AIDS Patients	Centers for Disease Control
Database on Homicides in the US	Federal Bureau of Investigation
Epidemiological Data on Air Pollution and Lung Function	University of Massachusetts Lowell Department of Work Environment
French Agricultural History	University of Massachusetts Lowell Department of History
Census Data on Engineers	US Census Bureau

Table 1. A list of databases used with our visualization approaches.

- Visual Programming Environment. This organization of Exvis' resources will allow a user to select the appropriate module for a particular task and incorporate it in a visualization pipeline, which provides file access, data manipulation, image processing, and other standard services.
- A Psychometric Module. To perform the perception experiments essential to our visualization research, we need a module that can display data pictures in any Exvis format, prompt the user for input, and accept and record responses.
- Support for Data Analysis. We need to build into the system the capability to identify a region of interest in the display and to compute various statistical measures of the data lying within the selected region. This will require convenient means for accessing powerful statistical analysis packages from within Exvis.
- Tools for Visualization of Large Databases. We are just beginning to address the problems involved in representing large databases (about 1,000,000 records). In addition to the obvious problems of storage and access, we are also concerned with finding effective ways to display datasets so large that each record cannot be represented by its own icon on the screen. We need to develop appropriate data filtering and combining operations, and panning and zooming techniques.
- 3D and Stereographic Displays. An important step toward perceptually compelling realism is to create icons in 3D and to display them in stereo.

4.2   Establish a Program of Research

To date we have devoted our energies mainly to explaining these new techniques in position papers and to generating illustrative displays. We have conducted little formal research, either to measure the effectiveness of these techniques in real-world applications or to increase our understanding of how they work and how to improve them. We need now to move on both fronts. Measuring how well these techniques work in real-world applications is, for any given case, a substantial undertaking. Such studies require access to large sets of well-documented data and usually involve two substantial stages of study. First, working closely with the scientists who generate the data, initial studies on one set of independent samples are required to establish what the displays can reveal and how best to tune and interact with them to bring out that information. Then, verification tests on yet other independent samples are required to measure precisely how effective the technique is. We have been working on pilot versions of such studies, and on proposals to agencies concerned with particular applications, to obtain funding for full-fledged studies. Obtaining fundamental knowledge of how these displays work and how to improve their effectiveness requires a considerably different form of research. In general one seeks to work with displays of artificial data, where one can specify and control precisely the kind and degree of structure present in the data samples. The effectiveness of our techniques for revealing these structures can then be precisely measured. Within a given technique, we can track the effects of systematic changes in particular display or interaction parameters, and begin to model how the technique works and how to maximize its effectiveness. We are strongly intent on setting up such studies. We also see the need to tap more effectively the knowledge and efforts of those doing basic research on the low-level perceptual processes we are trying to harness. Their understanding of these processes could help us to make more effective displays and guide us in our fundamental studies. Their awareness of what we are doing might also induce them to conduct studies more directly helpful to us.

References

1. R.M. Pickett. 1988. "Current Concepts: Person-Machine Systems for Imaged-Based Diagnosis," Investigative Radiology, 23(5), 394–395.

2. R.M. Pickett and G. Grinstein. 1988. "Iconographic Displays for Visualizing Multidimensional Data," Proceedings of IEEE International Conference on Systems, Man, and Cybernetics, Beijing and Shenyang, China, 514–519.

3. R.D. Bergeron and G. Grinstein. 1989. "The Impact of Scientific Visualization on Workstation Development," IFIP Workshop on Workstations for Experiments, Vol. 1, pp 3–11.

4. R.D. Bergeron and G. Grinstein. 1989. "A Reference Model for the Visualization of Multi-Dimensional Data." Eurographics ' 89, 393–399.

5. G. Grinstein, R.M. Pickett and M.G. Williams. 1989. "Exvis: An Exploratory Visualization Environment," Proceedings of Graphics Interface '89, 254–261.

6. S. Smith and M.G. Williams. 1989. "The Use of Sound in an Exploratory Visualization Environment," University of Lowell Computer Science Department Report No. 89-002.

7. M.G. Williams, S. Smith and G. Pecelli. 1989. "Computer-Human Interaction Issues in the Design of an Intelligent Workstation for Scientific Data Visualization (Phase 1)." ACM SIGCHI 1989 Bulletin.

8. G. Grinstein and S. Smith. 1990. "The Perceptualization of Scientific Data," Proceedings of the SPIE/SPSE Conference on Electronic Imaging, Extracting Meaning form Complex Data: Processing, Display, Interaction, Santa Clara, CA, Vol. 1259, 190–199.

9. H. Levkowitz and G.G. Grinstein. 1990. "Experimental Approaches to Color," Proceedings of the International Electronic Imaging Conference '90, Boston, MA, pp. 434–437, October 29–November 1, 1990.

10. H. Levkowitz and R.M. Pickett. 1990. "Iconographic Integrated Displays of Multiparameter Spatial Distributions," in B.E. Rogowitz and J.P. Allebach, eds., Human Vision and Electronic Imaging: Models, Methods, and Applications, February 12–14, Santa Clara, CA: SPIE, pp. 345–355.

11. R.M. Pickett, H. Levkowitz and S. Seltzer. 1990. "Iconographic Displays of Multiparameter and Multimodality Images," First Conference on Visualization in Biomedical Computing, Atlanta, GA, May 22–25, IEEE Computing Society Press, 58–65.

12. S. Smith. 1990. "The Perceptualization of Scientific Data." Video Supplement, Extracting Meaning from Complex Data: Processing, Display, Interaction, Vol. 1259-V, Bellingham, WA, SPIE, 1990.

13. S. Smith, R.D. Bergeron and G. Grinstein. 1990. "Stereophonic and Surface Sound Generation for Exploratory Data Analysis," CHI '90 Empowering People, Conference on Human Factors in Computing Systems, Seattle, Washington, April 1–5, Addison-Wesley (Proceedings), 125–132. In M. Blattner and R. Dannenberg, eds., Multimedia and Multimodal Interface Design, ACM Press.

14. H. Levkowitz. 1988. "Color in Computer Graphic Representation of Two-Dimensional Parameter Distributions," Ph.D. thesis, Department of Computer and Information Science, University of Pennsylvania, Philadelphia, PA. Technical Report MS-CIS-88-100, Department of Computer and Information Science and MIPG139, Medical Image Processing Group, Department of Radiology, University of Pennsylvania.

15. H. Levkowitz. 1991. "Color Icons: Merging Color and Texture Perception for Integrated Visualization of Multiple Parameters," in Visualization ' 91, San Diego, CA, October 22–25. IEEE Computer Society, IEEE Computer Society Press.

16. H. Levkowitz. 1991. "Exploratory Data Visualization: The Human Visual System should be the Main Design Consideration," Proceedings of the 1991 Joint Statistical Meetings—American Statistical Association (ASA), Atlanta, GA, August 18–22, 1991.

17. S. Smith. 1991. "Global Geometric, Sound, and Color Controls for the Visualization of Scientific Data." Video Supplement, Extracting Meaning from Complex Data: Processing, Display, Interaction. Vol. 1259-V, Bellingham, WA: SPIE.

18. S. Smith. 1991. "Interactive Data Exploration with a Supercomputer." Proceedings of Visualization '91—Video Journal, San Diego, CA, 1991.

19. S. Smith, G. Grinstein and R. Pickett. 1991. "Global Geometric, Sound and Color Controls for Iconographic Displays of Scientific Data," Proceedings of the SPIE/SPSE Conference on Electronic Imaging, Extracting Meaning from Complex Data: Processing, Display, Interaction, Santa Jose, CA.

20. S. Smith, G. Grinstein and D.R. Bergeron. 1991. "Interactive Data Exploration with a Supercomputer," Proceedings of Visualization '91, San Diego, CA.

21. S. Smith and L. Scarff 1992. "Combining visual and IR images for sensor fusion." Visual Data Interpretation, SPIE vol. 1668: pp. 102–112,.

22. K. Seetharaman, G. Grinstein, H. Levkowitz, and R.D. Bergeron. 1993. "A Conceptual Model for Interaction in Multiple Representational Spaces," In International Conference on Computer Graphics '93 Proceedings, Bombay, India, pp. 121–128, 1993.

23. K. Seetharaman, G. Grinstein, and H. Levkowitz. 1993. "Interactions in Color Space," In IS&T's 46th Annual Conference Proceedings, Boston, MA, pp. 134–135, 1993.

24. H. Levkowitz and G.T. Herman. 1993. "GLHS: A Generalized Lightness, Hue, and Saturation Color Model," CVGIP: Graphical Models and Image Processing, Vol. 55, No. 4, pp. 271–285, 1993.

25. K. Seetharaman. 1994. "An Interaction Model for Exploratory Data Visualization," Sc.D. Dissertation, Department of Computer Science, University of Massachusetts Lowell, Lowell, MA 1994.

26. Lee, J.P. 1994. "Data Exploration Interactions and the ExBase System," in Database Issues for Data Visualization, J.P. Lee and G.G. Grinstein eds., Lecture Notes in Computer Science, Springer Verlag, to appear, 1994.

27. S. Smith, H. Levkowitz, R. Pickett and M. Torpey. 1994. "A System for Psychometric Testing of Auditory Representations of Scientific Data." Proceedings ICAD '94, November 1994, Santa Fe, NM, to appear.

28. R.M. Pickett and B.W. White. 1966. "Constructing Data Pictures," Proceedings of the Society of Information Display Seventh National Symposium, Boston, MA, 75-81.

29. R.M. Pickett. 1968. "Perceiving Visual Texture: A Literature Survey," Aerospace Medical Research Laboratories Technical Report No. AMRL-TR-68-12.

30. R.M. Pickett. 1970. "Visual Analyses of Texture in the Detection and Recognition of Objects." in B.S. Lipkin and A. Rosenfeld, eds., Picture Processing and Psycho-Pictorics, New York: Academic Press, 189–208.

31. R.M. Pickett and D.J. Getty. 1981. "Demonstration of an Advanced Technique for Graphic Display of Multidimensional Data," BBN Report No. 4821.

An Environment and Studies for Exploring Auditory Representations of Multidimensional Data

Haim Levkowitz, Ronald M. Pickett, Stuart Smith, and Mark Torpey

Institute for Visualization and Perception Research
Department of Computer Science
University of Massachusetts Lowell
Lowell, MA 01854, USA

Abstract. The field of auditory data representation has produced several intriguing proof-of-concept systems, but up until now there has been little formal research to measure the effectiveness of auditory data displays or to increase our understanding of how they work and how to improve them. Formal assessment is necessary throughout the process of developing new auditory display technologies in order to learn how to restrict the universe of possible sound attributes to those that are most effective for data representation. The capability to run quick psychometric tests to obtain quantitative figures of merit for alternative auditory representations is a requirement for auditory-display researchers engaged in the development of new technologies. For the first time, this capability is realized with a special-purpose workstation designed to generate and administer psychometric tests automatically using test patterns generated from statistically well-specified synthetic data. We describe the characteristics of one such workstation we have developed. We also describe a testing methodology we propose for the development of new auditory data displays of a type that we have been working with for the last few years. Finally, we describe a specific set of studies we are now beginning to conduct.

1   Introduction

In the last decade we have seen a significant growth in the flow of scientific data. With it comes the need to improve the capabilities for exploratory data analysis. Techniques for data visualization offer great promise, but in spite of the many advances made, the ability of visualization to deal with high-dimensional data is still limited.

Investigators seeking ways to raise the dimensionality of visual data representations, or alternatives to visual display, have been attempting to develop effective methods for encoding quantitative information in sound [3, 24, 25, 28, 42, 45, 55]. In experiments comparing subjects' performance on a variety of data analysis tasks using visual data representations alone and combined visual and auditory data representations, subjects have shown modestly improved performance when using the combined representations (see, for example, [3, 16, 28, 54]). While these studies have proved the concept of auditory data representation, much work remains to be done to establish the value of, and effective approaches for, auditory data representation as a tool for analyzing and exploring data.

Up until now there has been little formal research, either to measure the effectiveness of auditory data displays in real-world applications or to increase our understanding of how they work and how to improve them. Frysinger [9, p. 136] has given a concise statement of the overall task that must be accomplished. He states the need to discover the set of useful auditory parameters and to understand their perceptual transfer functions. With such knowledge, it will be possible to design displays that take advantage of the most useful parameters. Also, we need to find those data analysis tasks that can most benefit from auditory data representation, and the types of displays that would be best to apply to them. Finally, the interactions between visual and auditory data representations have to be understood to be able to choose the best combination of the two for a given analysis task.

1.1 Sound generation environments

Systematic and quantitative studies of auditory data representation require significant capabilities for precise specification and manipulation of sounds. In particular, it is necessary to have sound generation techniques that offer rich sets of predictable, perceptually relevant transformations. These transformations allow data dimensions to be associated with perceived attributes of sound. Among the many potentially useful sound synthesis methods are FM [5, 6], non-linear distortion or "waveshaping" [17], and granular synthesis [37, 38].

Smith et al. [46] outline the characteristics of a sound facility for auditory data representation. We briefly summarize it here. A sound facility for auditory data representation should offer the ability to synthesize complex sounds in real time and the ability to construct arbitrary sound-synthesis software modules in a convenient way. The facility needs to have both special-purpose sound-generating hardware and special-purpose software that offers (1) powerful sound description capabilities, (2) the control and timing functions necessary to cause each sound to happen at a precisely specified time, and (3) a graphical user interface. Recently, two systems offering such capabilities have appeared: the IRCAM Music Workstation [21, 22, 34, 35, 51], which is built around the NeXT computer, and the Kyma system [40, 41], which uses a Macintosh II or 486 PC and a special attached sound processor. While neither of these systems was designed specifically for auditory data representation, both are adaptable to this purpose.

In Section 2, we survey and discuss various approaches to representing data in sound. In Section 3, we describe the system and environment we have developed to generate data-controlled sounds. In Section 4, we discuss in general the need to provide efficient evaluation procedures for comparing alternative auditory representations. We describe the components of such studies, and present a program of studies we are about to conduct. We also describe briefly the development environment we have developed, which is necessary for designing, developing, and conducting such studies. Section 5 summarize and concludes with the expectations for better understood auditory data displays upon the conclusion of our studies in Winter 1994.

2 Representing data in sound

Consider a large collection of multidimensional data in which each data item comprises r individual values, called its dimensions. The creation of an auditory representation of such data is perhaps best accomplished via a two-level process. At the first level, an auditory element is created from each data item. At the second, a display is created from a whole collection of such elements.

2.1 Auditory elements

There are many design options for auditory elements. A fundamental option is whether to create natural-sounding elements (like those emitted by real physical objects and events) or arbitrary abstract ones. While natural sounds are generally more difficult to implement computationally, they may evoke more effective perceptual analysis of the data than abstract sounds (for the key issues concerning the synthesis of natural sounds for computer interfaces see [10]).

Each auditory element has distinctive audible attributes that are controlled by the data item's values. A multidimensional data item can be mapped onto such an element in many different ways. Its dimensions can be mapped, for example, to pitch, loudness, duration, attack and decay rates, the parameters of amplitude and frequency modulation, the parameters of various kinds of filters, rates of pitch slide (*glissando* or *portamento*) and changes in waveform. For n sound attributes and r data dimensions ($r \leq n$) the number of different ways to pair data dimensions with sound attributes is the number of permutations $P_r^n = n!/(n - r)!$. Even for n and r as small as 10 there are well over three *million* possible ways to pair sound attributes and data dimensions. The many different mapping functions that can be used for each pair increase the number of possibilities even further.

Current knowledge of human auditory perception does not provide those combinations of sound attributes that are right for the representation of multidimensional data. Most of the basic research on audition has focused on the perception of single sound attributes [43, 44] or interactions between two attributes [7, 8, 11, 31, 36, 48, 49, 50, 56]. Investigations of the perception of musical timbre [2, 12, 13, 32] have attempted to find a small number of auditory dimensions on which differences in the tonal qualities of sounds can be represented; however, these efforts have not produced a satisfying comprehensive theory of timbre perception. In vision, on the other hand, we have models that are well understood. For example, color models can be divided into three types: process dependent—e.g., Red, Green, Blue (RGB); Cyan, Magenta, Yellow (CMY)—whose coordinates describe processes or devices; pseudo-perceptual—e.g., Lightness, Hue, Saturation (LHS)—whose coordinates describe perceptual dimensions of color but are not perceptually calibrated; and perceptual/uniform—e.g., Munsell, CIELUV—whose coordinates are obtained from empirical measurements of perceptual dimensions. All are based on known definitions, and possess properties that are psychologically understood [19]. There are no timbre models corresponding to any of these color models; timbre dimensions do not behave in psychologically understood ways and are not known to correspond well to physical definitions.

2.2 Auditory data displays

At the second level of designing an auditory representation, we are concerned with how to display a collection of auditory elements in a sufficiently compact way to evoke perception of the characteristics (such as statistics) of the data. One basic approach is to present the elements in rapid sequence; another is to present them in parallel, as a kind of ensemble; a third is to create ensemble sequences. As long as implementation is feasible, perceptual considerations should have the highest priority.

We suggest as a starting approach to auditory data representation a method that is an extension of an "iconographic" approach we have been developing for data visualization [14, 18, 30]. In our visual iconographic approach, each data item is represented by an icon whose visual features (such as orientation, thickness, or length) and attributes (such as color) are controlled by the data. Data can be mapped to features and attributes. Two of the data variables, not necessarily independent of those controlling the icon features and attributes, control the position of each icon on the display surface. With sufficient density, the icons form a surface texture display, and structures in the data are revealed as different texture regions.

We have used the success of displays based on visual texture perception as a lead for the development of data representations based on the perception of auditory texture, see, e.g., [46]. By sweeping the mouse cursor over an iconographic visual display, the user can trigger an auditory data representation consisting of multiple small and partially overlapping sonic events, such as tones or noise bursts. The textural character of the sound produced can change audibly when the characteristics of the data controlling the sounds change, thereby indicating the presence of a boundary or contour, which can be seen visually in the image.

The idea for using sound in this way finds some support in the work of Bregman. Bregman [4] has speculated on the existence of *auditory* grain analyzers capable of segregating "streams" of sound on the basis of texture differences. Bregman defines an auditory stream as the perceptual unit that represents a single happening [4, p. 10]. The stream plays the same role in auditory mental experience that the object plays in visual experience. Bregman cites an experiment by Warren and Verbrugge [52] that shows that the auditory system, essentially on the basis of texture differences, can easily distinguish the sound of a bottle bouncing on a resilient surface from that of a bottle breaking against a hard surface. Bregman [4, p. 121] concludes that "a granular analysis might provide a perceptually relevant description of the sounds of events that can be considered iterations of multiple small events."

Bregman cautions [4, p. 121] that it may not be practical to use granularity statistics to create a *sharp* boundary between streams because the auditory system might need a temporally extended sample in order to assess the statistical properties of each granularity. We do not consider this a serious problem for our approach because we use sound not alone, but in support of visual analysis. Therefore, it is not necessary that sound establish a sharp boundary between regions. It is sufficient if sound can call attention to a boundary and indicate its approximate location for subsequent visual and conventional mathematical analysis.

3 Generating data-controlled sounds

3.1 Hardware and software for data-controlled sound

We have developed a prototype environment that can support research in auditory data representation. This environment is based on a configuration of an Apple Macintosh computer and the KYMA sound generation and manipulation system [40, 41]. KYMA uses the Macintosh to provide all required computing functions except real-time control and sound generation/processing. For the latter, KYMA employs a special attached processor that connects directly to the Macintosh bus.

3.2 Hierarchical construction of sounds

The sounds produced by our auditory data representation facility has been constructed hierarchically. The hierarchy has three levels:

Grain: the fundamental unit, or primitive, of our auditory data representation. A grain is a brief sound having a standard amplitude envelope and fixed frequency and waveform.

Event: an assemblage of grains forming a temporally extended unit such as a tone or noise burst. An event's acoustic properties can change over time. For example, an event might be a tone whose pitch slides up and down like a siren.

Stream: an assemblage of events having similar or logically related auditory characteristics. This usage of the term "stream" is intended to parallel Bregman's definition of a stream as the perceptual unit that represents a single happening.

Data control is applied at the level of the *event*. Data do not directly control the frequencies, amplitudes, and waveforms of individual grains. Data are instead treated as the values of the global control parameters of various event-generating routines. These routines produce the aggregations of grains necessary to create the desired sound. For example, a routine to generate a tone that slides upward from a given pitch at a specified rate would generate a sequence of grains, each having a frequency higher by some factor than its predecessor's.

3.3 Interactive capabilities

3.3.1 Display designer interaction Our system provides the designer of auditory data displays with a set of interactive software tools that permit rapid prototyping of displays having different characteristics. This tool set is organized according to the grain/event/stream hierarchy explained above, i.e., it provides tools at the grain, event, and stream levels. Grain-level tools generate individual grains. Among the grain-level tools are the simple grain generator described by Roads [37]. Event-level tools include modules that generate multiple grains on command. At this level, the designer is concerned with producing a sound with particular timbral qualities and does not necessarily know or care precisely how many grains will be generated by an event. Event-level tools provide global control over such attributes as the number of grains generated per unit of time and the distribution of the frequencies of the grains. Stream-level tools are concerned primarily with determining the type and degree of interaction provided to the system user. Modules that run particular types of psychometric tests are stream-level tools.

3.3.2 User interaction The auditory data display is always in the form of one or more streams of events. The component events of a stream may be sequential or overlapping, and two or more streams may exist simultaneously. Streams can be produced either by user interaction with a display, by an automatic mechanism that plays collections of events according to some set of rules, or by some combination of user interaction and automation. We built into the system two general display modes—one in which the user is a passive listener to streams, where the flow is controlled by the system, and the other where the user can control the flow, including replaying selected portions of the stream.

4 Evaluation Methods and Procedures

We are about to start our studies for evaluation of auditory data displays. Our main approach to evaluating the different auditory representation schemes that we develop is to run psychometric tests to obtain a quantitative figure of merit for each scheme. These tests employ test patterns generated from statistically well-specified synthetic data. Supplementing these quantitative evaluations will be other qualitative assessments of how effective these representation schemes are in exploring structure in real data.

At this early stage in the development of the field of auditory data representation, we are concerned not only with determining how well particular representations reveal structures of particular kinds in data, but also with understanding *how* these representations work. We need this knowledge to steer our explorations of the large space of possible representations in order to converge on the optimal ones. Gathering this knowledge requires systematic and quantitative testing of auditory representations, using test patterns that contain well-specified and systematically manipulable structures.

Like designing auditory representations of data, evaluating representations confronts the investigator with many options. There are two general considerations: the type of structure in the data to study and the kind of auditory task to use to study it. We first examine the structure question, followed by a brief discussion of auditory tasks.

4.1 Structures in data

Real-world data. In choosing structures, one direction to look is toward the real world settings in which the representations are expected to be employed; however, the common problem in exploring real data is that more often than not the structures in such data are not well understood. If they were well understood, it would not be important to study them with the techniques discussed here. Also, there is usually little available truth data (i.e., information independent of the database that establishes the presence, location, and nature of a known structure) on which to select reliable test patterns. Ultimately, a representation should be validated with real data, but it is reasonable to begin by looking at fully specified and systematically controlled data structures embedded in *synthetic* data.

Synthetic data. A synthetic database can be created by some random process. The output of the process can be mapped to some spatial or temporal interval of the display and onto attributes of the elements displayed at each location within the interval. Structures are created by altering one or more parameters of the process as it is mapped over the display interval. In this way one can create sets of test patterns in which the type of structure, its location in the data base, and other characteristics can be precisely specified and controlled. The challenge for the researcher is to make good choices of the process, the parameters to be altered, and the patterns of alteration over the display interval.

It is reasonable to work first with a familiar and well understood random process, like a multidimensional Gaussian, and with differences in a simple statistic like the mean or variance in one or more dimensions. One might then manipulate also correlations among the dimensions. The random process samples from a Gaussian-shaped cloud in n-space. Differences in the mean of one or more dimensions move the cloud around in n-space, while differences in the variance of one or more dimensions elongate or compress the cloud along those dimensions. Differences in correlation change the cloud's shape and orientation. Working within this concept of sampling clouds of points in n-space, Bly [3] created test patterns by sampling points from cylindrical structures. But the world of possibilities is mostly beyond our current understanding. There is a critical need for help, possibly from the field of statistical pattern recognition.

4.2 Tasks

Three main types of tasks are used in studies of this kind: detection, recognition, and quantification.

Detection. Detection is the simplest task and the one we will be using in the starting phase of research. The listener is given a series of test patterns that contain a difference in the structure of interest, randomly interleaved with test patterns that do not contain a difference. The listener's task is to identify the patterns that contain a difference. The listener's performance provides objective and quantifiable evidence of how effective a given representation is in making the difference audible.

Detection studies are a very powerful form of evaluation; however, they only provide information about the listener's ability to hear a difference, not about the nature of the difference itself. That kind of information has to come from studies using recognition and quantification.

Recognition. A recognition task is one in which the listener is asked to demonstrate not just that the difference is audible, but also that it is audibly discriminable from one or more other kinds of differences.

Quantification. Here the listener provides quantitative information about the difference.

4.3 The studies

A full program of research should be aimed not just at measuring effectiveness, but also at understanding how the representation works perceptually, and therefore how to improve it. Such a program would require studies using all three tasks. We briefly describe a starting approach to evaluation, limited to a particular type of detection task and to simple structures embedded in synthetic data.

The general approach is to create test patterns consisting of streams of auditory events in which each event is an auditory rendition of a data item from a normally distributed random generator. A two-alternative forced choice testing procedure is employed; for each test trial two such streams are presented in succession. One of the streams, chosen at random, contains a midway difference in its statistical characteristics. The differences can be in the mean of one or more of the dimensions controlling the attributes of the events, or in the correlation among two or more of the dimensions. The listener has to identify which of the two streams contains the difference.

The test trials follow an adaptive up-and-down testing procedure [20] in which the size of the difference is varied from trial to trial depending on the listener's performance. On the first trial, the size of the difference is set to be so large that the listener cannot fail to hear it. In subsequent trials, the size of the difference is decremented until the listener begins to make mistakes in choosing the stream that contains the difference. The size of the difference is then incremented in subsequent trials until the listener demonstrates reliably correct choices. From an averaging of the degree of difference at which each reversal occurs, a threshold value can be computed. These threshold values can then serve as quantitative figures of merit for comparing discrimination performance across different data representations and across individual attributes or sets of attributes within a single data representation.

4.4 Development environment

Smith et al. [46] have presented the specifications of an environment for developing and studying auditory representations of data. Such an environment must provide tools for running formal and informal experimental studies, like those described above. The environment must automate routine activities, such as creating test patterns, presenting the test patterns to human test subjects, recording subjects' responses, and analyzing the recorded data. It must also provide support for non-routine activities, such as designing the parameters for an experiment. We have developed an environment that supports all these requirements.

5 Summary and Conclusions

The increasing power of sound generation and manipulation systems, presents auditory-display researchers with the need to consider a vastly expanded universe of possible auditory data representations. This universe is so large that researchers need tools that allow quick pruning of unpromising representations and permit the exploration of alternative representations to converge rapidly and efficiently on those that are most effective. We have described the various aspects of an environment that we have developed, which supports systematic and quantitative testing of a wide range of auditory data representations. The heart of this environment is a sophisticated sound generation system and a toolkit for creating test patterns that contain well-specified and systematically manipulable data structures. This environment provides tools for running formal and informal experimental studies. The environment also provides tools to keep track of the path and results of exploration so that it is possible to build systematic knowledge and theory in this new and poorly understood area. We have also described a set of studies we are about to start conducting to determine optimal auditory data representation techniques. When completed (around Winter 1994), these studies will provide the knowledge necessary to select an optimal data representation for a given task.

Acknowledgments

This research is supported by NSF Grant No. IRI-9214694.

References

1. Bartram, L., K. Booth, and W. Cowan. "Issues in the Design of Workstations for Psychology Experimentation." in J. Encarnacão and G. Grinstein (eds.) *Workstations for Experiments.* Berlin: Springer-Verlag, 1991.

2. von Bismarck, G. "Timbre of steady-state sounds: a factorial investigation of its verbal attributes," *Acustica,* (30): 146-159, 1974.

3. Bly, S. "Presenting information in sound." *Proceedings of the CHI '82 Conference on Human Factors in Computer Systems,* 371-375, 1982.

4. Bregman, A. *Auditory Scene Analysis.* Cambridge, MA: MIT Press, 1990.

5. Chowning, J. "The Synthesis of Complex Audio Spectra by Means of Frequency Modulation." *Journal of the Audio Engineering Society,* 21(7):526-534, 1973.

6. Chowning, J., and D. Bristow. *FM Theory and Applications.* Tokyo: Yamaha Music Foundation, 1986.

7. Doughty, J., and W. Garner. "Pitch characteristics of short tones II: pitch as a function of duration." *Journal of Experimental Psychology,* 38:478-494, 1948.

8. Fletcher, H., and W. Munson. "Loudness, its definition, measurement and calculation." *Journal of the Acoustical Society of America,* 5:82-108, 1933.

9. Frysinger, S. "Applied research in auditory data representation." *Proceedings of the SPIE/SPSE Conference on Electronic Imaging,* 1259:130-139, 1990.

10. Gaver, W. "Synthesizing Auditory Icons." *Proceedings of the International Conference on Auditory Display (ICAD),* Reading, MA: Addison-Wesley, 1994.

11. Green, D., T. G. Birdsall, and W. P. Tanner. "Signal detection as a function of intensity and duration." *Journal of the Acoustical Society of America,* 29:523-531, 1957.

12. Grey, J. *An exploration of musical timbre.* Stanford, CA: CCRMA/Stanford Department of Music, 1975.

13. Grey, J. "Multidimensional perceptual scaling of musical timbres." *Journal of the Acoustical Society of America,* 61:1270-1277, 1977.

14. Grinstein, G., and S. Smith. "The perceptualization of scientific data." *Proceedings of the SPIE/SPSE Conference on Electronic Imaging,* 1259:190-199, 1990.

15. Howe, H. *Electronic Music Synthesis.* New York: W. W. Norton, 1975.

16. Fitch, T. and G. Kramer. "Sonifying the Body Electric: Superiority of an Auditory over a Visual Display in a Complex, Multivariate System." *Proceedings of the International Conference on Auditory Display (ICAD),* Reading, MA: Addison-Wesley, 1994.

17. Le Brun, M. "Digital waveshaping synthesis." *Journal of the Audio Engineering Society,* 27(4) :250-266, 1979.

18. Levkowitz, H. "Color icons: Merging color and texture perception for integrated visualization of multiple parameters." In *Visualization '91,* San Diego, CA, October 22–25 1991. IEEE Computer Society, IEEE Computer Society Press.

19. Levkowitz, H. and G. T. Herman. "GLHS: A generalized lightness, hue, and saturation color model." *CVGIP: Graphical Models and Image Processing,* 55(4):271–285, 1993.

20. Levitt, H. "Transformed up-down methods in psychoacoustics." *Journal of the Acoustical Society of America,* 49, 1972.

21. Lindemann, E., F. Dechelle, B. Smith, and M. Starkier. "The architecture of the IRCAM Musical Workstation." *Computer Music Journal,* 15(3): 41-49, 1991.

22. Lindemann, E., and M. de Cecco. "Animal: graphical data definition and manipulation in real time." *Computer Music Journal,* 15(3): 78-100, 1991.

23. Lunney, D., and R. Morrison. "High technology laboratory aids for visually handicapped chemistry students." *Journal of Chemical Education,* 58(3):228-231, 1981.

24. Lunney, D., and R. Morrison. "Auditory presentation of experimental data." *Proceedings of the SPIE/SPSE Conference on Electronic Imaging,* 1259:140-146, 1990.

25. Mansur, D., M. Blattner, and K. Joy. "Sound graphs: a numerical data analysis method for the blind." *Journal of Medical Systems,* 9(3):163-174, 1985.

26. Mathews, M. *The Technology of Computer Music.* Cambridge, MA: MIT Press, 1969.

27. Mathews, M., and J. Pierce, eds. *Current Directions in Computer Music Research.* Cambridge, MA: MIT Press, 1989.

28. Mezrich, J., S. Frysinger, and R. Slivjanovski. "Dynamic representation of multivariate time series data." *Journal of the American Statistical Association,* 79(385):34-40, 1984.

29. Moore, F. *Elements of Computer Music.* Englewood Cliffs, NJ: Prentice-Hall, 1990.

30. Pickett, R., and G. Grinstein. "Iconographic displays for visualizing multidimensional data." *Proceedings of the 1988 IEEE Conference on Systems, Man and Cybernetics.* Beijing and Shenyang, People's Republic of China, 1988.

31. Plomp, R., and M. Bouman. "Relation between hearing threshold and duration of pulses." *Journal of the Acoustical Society of America,* 31:749-758, 1959.

32. Plomp, R. "Timbre as a multidimensional attribute of complex tones," in R. Plomp and G. F. Smoorenburg, eds., *Frequency Analysis and Periodicity Detection in Hearing,* Leiden, Netherlands: Sijthoff, 1970.

33. Pollack, I., and L. Ficks. "Information of elementary multidimensional auditory displays." *Journal of the Acoustical Society of America,* 26:155-158, 1954.

34. Puckette, M. "FTS: a real-time monitor for multiprocessor music synthesis." *Computer Music Journal,* 15(3): 58-67, 1991.

35. Puckette, M. "Combining event and signal processing in the MAX graphical programming environment. *Computer Music Journal,* 15(3): 68-77, 1991.

36. Reichardt, W., and H. Niese. "Choice of sound duration and silent interval for test and comparison signals in the subjective measurement of loudness." *Journal of the Acoustical Society of America,* 47:1083-1090, 1970.

37. Roads, C. "Granular synthesis of sound." *Computer Music Journal,* 2(2):61-61,1978.

38. Roads, C. "Asynchronous granular synthesis," in G. De Poli, A. Piccialli, and C. Roads eds. *Representations of Musical Signals.* Cambridge: MIT Press, 1991.

39. Roads, C., and J. Strawn, eds. *Foundations of Computer Music.* Cambridge, MA: MIT Press, 1985.

40. Scaletti, C. "The Kyma/Platypus computer music workstation," in S. Pope ed. *The Well-Tempered Object: Musical Applications of Object-Oriented Software Technology.* Cambridge: MIT Press, 1991.

41. Scaletti, C., and K. Hebel "An object-based representation for digital audio signals," in G. De Poli, A. Piccialli, and C. Roads eds. *Representations of Musical Signals.* Cambridge: MIT Press, 1991.

42. Scaletti, C. "Using sound to extract meaning from complex data." *Proceedings of the SPIE/SPSE Conference on Electronic Imaging,* Vol. 1459:207-219, 1991.

43. Scharf, B., and S. Buus. "Audition I: stimulus, physiology, thresholds." in K. R. Boff, L. Kaufman, and J. P. Thomas (eds.) *Handbook of perception and human performance,* 1:14.1-14.71). New York: Wiley, 1986.

44. Scharf, B., and A. Houtsma. "Audition II: loudness, pitch, localization, distortion, pathology." in K. R. Boff, L. Kaufman, and J. P. Thomas (eds.) *Handbook of perception and human performance* 1:15.1-15.60). New York: Wiley, 1986.

45. Smith, S., R. Bergeron, and G. Grinstein. "Stereophonic and surface sound generation for exploratory data analysis." *Proceedings of CHI '90,* Seattle, WA, 1990.

46. Smith, S., R. M. Pickett, and Marian G. Williams. "Environments for Exploring Auditory Representations of Multidimensional Data." *Proceedings of the International Conference on Auditory Display (ICAD)*, Reading, MA: Addison-Wesley, 1994.

47. Speeth, S. "Seismometer sounds." *Journal of the Acoustical Society of America*, 33:909-916, 1961.

48. Stevens, S. "The Relation of pitch to intensity." *Journal of the Acoustical Society of America*, (6):150-154, 1935.

49. Terhardt, E. "Pitch of pure of tones: its relation to intensity." in E. Zwicker and E. Terhardt (eds.) *Facts and Models in Hearing.* New York: Springer-Verlag, 1974.

50. Verschuure, J., and A. A. van Meeteren. "The effect of intensity on pitch." *Acustica*, 32:33-44, 1975.

51. Viara, E. "CPOS: a real-time operating system for the IRCAM Musical Workstation." *Computer Music Journal*, 15(3): 50-57, 1991.

52. Warren, W., and R. Verbrugge. "Auditory perception of bouncing and breaking events: a case study in ecological acoustics." *Journal of Experimental Psychology*, (10)5:704-712.

53. Williams, M. *Interactive Assistance for Experimentation on the Visual and Auditory Properties of Iconographic Data Displays.* Dissertation, University of Massachusetts Lowell, 1992.

54. Williams, M., S. Smith, and G. Pecelli. "Experimentally driven visual language design: texture perception experiments for iconographic displays." *Proceedings of the IEEE 1989 Visual Languages Workshop*, Rome, Italy, pp. 62-67.

55. Yeung, E. "Pattern recognition by audio representation of multivariate analytical data." *Analytical Chemistry*, (52)7:1120-1123, 1980.

56. Zwislocki, J. "Temporal summation of loudness." *Journal of the Acoustical Society of America*, 46:413-441, 1969.

Perceptual Principles for Effective Visualizations

Penny Rheingans   Chris Landreth

Martin Marietta Services Group
US EPA Visualization Center   MCNC
RTP NC, 27709   RTP NC, 27709
rheingans@vislab.epa.gov   landreth@ncsc.org

Abstract. Since visual data representations are perceived through the filter of the human visual system, it is imperative that the characteristics of this system be taken into account during the design and rendering of visual displays. This paper presents a set of perceptual guidelines for the construction of effective visualizations. In most cases, side-by side pictures demonstrate the impact of the suggested techniques.

1   Introduction

Visualization is the science of representing data visually in order to enhance communication or understanding. In this way, the complex and powerful capabilities of the human visual system can be harnessed to aid in the comprehension of potentially huge quantities of information. Since the information contained in visualizations must pass through the perceptual system, careful attention to the system's characteristics can greatly improve the effectiveness of visualizations.

This paper presents a number of perceptual principles for the construction of effective visualizations. Principles are demonstrated in example images. In the example images, value is represented by color, height, position, or opacity. Other representation parameters are certainly possible, but are not covered here. Additional cues which can enhance comprehension are also spotlighted.

2   Single-Variable Visualizations

The simplest class of visualizations show the value of a single scalar variable over a domain. The purpose of such images is usually to show the pattern of data values or the locations of data features (such as extreme values) in relation to geographic landmarks (such as state boundaries or anatomical features). In addition to such qualitative information, quantitative information about values at particular locations may be of interest.

2.1   Exploit familiar scenarios

Visualizations are most effective when their layout, lighting and choreography contain visual elements that are associated with common perceptual experience. Previous exposure to similar conditions lays the groundwork for a quicker and deeper comprehension of a visualization's geometric features.

Lighting. Through repeated experience, humans have become accustomed to illumination sources which originate from above the objects in a scene. For example, we usually view scenes illuminated by the sun or overhead light fixtures. Ramachandran [88] found that when subjects viewed images of objects that were ambiguously either concave or convex, the subjects always resolved the ambiguity based on the perceived direction of the light source from the objects.

Figure 1 illustrates the point. Two plates are displayed with circular objects visible within their bounds. The illusion is that the circular objects on the left plate appear as mounds, while the objects on the right appear as indentations. In fact, all the circular objects are mounds; the lighting on the right plate is from below. In general, lighting of complex 3D objects from below may cause similar confusion in the perception of those objects and should therefore be avoided.

Shadows. Displaying a 3D object in an environment where its shadow projects onto nearby flat surfaces is a powerful method of conveying three-dimensional structure and placement. Conversely, when shadows are absent, the object's structure or placement within the environment may be significantly misinterpreted. Figure 2 shows a comparison between a 3D isosurface rendered without (Fig. 2a) and with (Fig. 2b) a projected soft shadow onto an adjacent flat surface. Viewers who observed the scene without shadows were generally unable to conclude where the isosurfaces were located with respect to the map surface, but they could readily locate the same isosurfaces, when accompanied by shadows, as residing directly on top of the map surface.

Other cues. Some other shape and position cues of the physical world that can be effectively used in visualizations include: hidden line and surface removal, perspective, intensity depth-cuing, and stereo display. Hidden line and surface removal gives cues to the relative distances to objects along the same line of sight because nearer objects obscure farther ones. Perspective, intensity depth-cuing, and stereopsis all give depth cues.

2.2   Emphasize the interesting

Designers of visualizations should take care that the most striking features of the image are also the most important. Representations which draw the viewer's eye to unimportant features may cause more interesting features to be overlooked. Features likely to catch the eye are those that are brightly colored, moving or changing, defined by sharp boundaries, or highly saturated.

The common spectrum color scale maps the middle values to yellow, a particularly striking color. In applications where the location of middle values is of particular interest, this is appropriate. Such applications are not very common, however. More often, the high or low values are of greatest interest, and middle values are of least interest.

Double-ended Color Scales. Data sets with both positive and negative values can have a zero point representing no change, average, or expected value. In such data, deviation from zero (and the pattern of such deviation) is what is interesting. Figure 3 shows such a data set. Positive values show deposition of sediment from a column of water. Negative values indicate erosion and resuspension of sediment back into the water. Figure 3a uses a standard spectrum color scale to display the data. The distribution of erosion and deposition is not immediately obvious. A conscious distinction must be drawn between orange and red (indicating deposition) and the other hues (indicating erosion). Figure 3b shows the same data mapped with a double-ended color scale. In this image, the areas characterized by erosion and deposition are clearly and immediately defined. Areas with no change in sediment are an unobtrusive grey.

The concept of double-ended color scales extends naturally to bivariate color scales, i.e., mappings from two scalar values to a color. One such scale might map the value of one variable to brightnesses of a hue, such as green. The value of the other variable would be mapped to brightnesses of the complementary hue, in this case purple. The contributions of the two variables are summed additively to give the display color. The resulting color scale contains three clearly discernible classes of colors. Greys represent places where the values of the two variables are comparable because equal chromatic contributions of the two hues cancel each other, producing grey. Dark greys are formed when both values are low, while light greys result when both values are high. Places where one variable is significantly larger are colored green, while places where the other predominates are colored purple. Such a color scale is useful in situations where the two variables are expected to be correlated. Places where this relationship does not hold are immediately apparent.

Missing Values. Sometimes things that are not shown can be distracting. Figure 4a shows a height-mapped, pseudo-colored representation of ozone concentrations over the southern hemisphere. Places where no data value was available are made invisible, or *iblanked*. While iblanking seems to be the most accurate way to represent missing values, the sharp boundaries of the holes and the very different values peeking through them draw a viewer's eyes to the holes and away from the values that are actually present. The effect is even stronger when the visualization is animated to show changes in the distribution over time. In fact, when the visualizer showed this representation to the atmospheric scientists studying the data, they promptly asked that the holes be "filled in." Figure 4b shows the same data set with missing values estimated by an adaptive filter. No valid values were changed.

Interpolation is not without dangers, however. A visualization showing interpolated values can misrepresent the smoothness of data values, the density of data values, and the likelihood that displayed values are accurate. Ideally, both the interpolated and iblanked visualizations would be available to the researcher. Additionally, a visualization that mapped value to color and/or height and mapped certainty about estimated values (related inversely to distance from real data values) to opacity would be valuable.

2.3 Say it again (Use redundant mappings)

Visualizations that represent data values using multiple display parameters have the potential to portray the data more effectively than visualizations that map each data variable to a single display parameter. There are a number of compelling reasons why this should be true:

1. Different display parameters convey different types of information most efficiently. For example, brightness conveys shape more effectively than hue, but hue provides more accurately distinguishable display levels than does brightness.

2. Multiple display parameters can overcome visual deficiencies. If one display parameter is ambiguous or ineffective because of the visual deficiency, another may compensate. For example, a person with color deficiencies would likely find it easier to unambiguously judge the value represented by a color using a redundant hue and brightness color scale than one using a standard spectrum scale which varies only in hue.

3. Multiple display parameters reinforce each other. In this way, areas with differing values have greater visual difference from one another.

Color and Height. For example, Figure 5b shows a representation in which concentrations of hydrogen peroxide (H_2O_2) have been mapped to both height and color. This redundant representation conveys the shape of the statistical surface more clearly than one mapping values to color (Figure 5a), because humans are more experienced judging the pattern of a surface from its shape than from its color. Additionally, the redundant representation conveys the location of global maxima (or near maxima) more clearly than a representation using just height-mapping (Figure 5c). By using a redundant representation, different perceptual channels can simultaneously process features of the data distribution.

Redundant Color Scales. Redundant representations need not employ parameters as distinct as height and color. Values can be represented redundantly using only color. For example, data values can be mapped to both hue and lightness. Figure 6b shows such a visualization. Compared to an image using only hue (Figure 6a), this representation shows the location and swirling structure of high areas more clearly. This redundant representation also has the advantage that it can be unambiguously interpreted by someone with a dichromatic color deficiency. For example, a protanope (commonly called red-green color blind) viewing Figure 6a would find it difficult to distinguish the values near 100 (green) from the values near 170 (red). The same person viewing Figure 6b would be able to distinguish the two values based on lightness; the higher values would appear brighter. Because a legend is included, the value encoded by each color is unambiguous (to the limits of discrimination).

The utility of redundant color scales has been empirically confirmed. Ware [88] conducted three experiments comparing a linear grey scale, a perceptual grey scale, a saturation scale, a spectrum scale, and a red-to-green scale for univariate data representation. In the first experiment, subjects were asked to judge the metric value of a colored patch surrounded by a contrasting area. The spectrum scale produced significantly more accurate metric value readings. In the second experiment, subjects were asked to judge the effectiveness of the color scales in revealing information about the surface properties of simulated surfaces. In general, the grey scales were judged to be more effective. In the third experiment, the five original color scales were compared to a redundant experimental scale that cycled through the hues while it increased monotonically in lightness (a rainbow scale similar to that in Figure 6b). When the experimental scale was used for a metric query task, subject accuracy was similar to that of the spectrum scale (which had no monotonic lightness variation) and significantly better than the others. This suggests that a color scale that varies in both luminance and hue can be used to accurately represent both metric and surface properties.

Other redundant color scales are the heated-object [Pizer and Zimmerman 83] and optimal color scales [Levkowitz 88]. The heated-object scale goes from black through red, orange, and yellow to white, with brightness increasing monotonically. The heated-object scale has more distinguishable display values and more contrast between different levels than a grey scale [Pizer and Zimmerman 83]. The heated-object scale has a stronger perceived natural ordering than the rainbow scale because of the monotonic increase in brightness and because the color order is based on experience. Levkowitz's optimal color scales increased monotonically in both brightness and RGB components while being linearized with respect to just noticeable differences (JNDs). Levkowitz experimentally compared the optimal color scale to a linearized grey scale and a linearized heated-object scale. He found the grey scale to result in significantly more accurate identifications of simulated lesions in medical images. While this result does not exactly validate the use of redundant color scales, the expect results of the experimental task (shape perception) was that the grey scale would excel.

Explicit Redundancy. In contrast to mapping a given data field to a single object having redundant attributes (e.g. color and height), the same data field may be mapped to two or more redundant objects, each having a single distinct attribute. Figure 7a shows a visualization in which the surface of the planet Jupiter has been mapped to two images, one with true color and one with a hue/value pseudo-colored field. With multiple representations such as this, accuracy in metric readings (pseudo-colored image) is maximized while maintaining optimal clarity of the object's form (true color image). In Figure 7b, a single molecule is redundantly displayed four times: as a single solid object, and as three orthogonal projections of the solid object. The object's form is best revealed by the solid representation, while quantitative information of the object's geometry in Cartesian space is revealed by the orthogonal projections. In both examples, redundancy is used to eliminate the need to compromise mutually exclusive visual elements of a single representation by allowing simultaneous display of all elements.

More Redundant Techniques. Other examples of redundant representations include combining size and color, opacity and color, and color and texture. A vector data set can be represented using both color and vector length to redundantly encode vector magnitude. Color and opacity can be used redundantly to represent value, mapping more important values to higher opacity so that interesting places are more apparent.

2.4   Minimize illusions

The human visual system is not immune to confusion. Specifically, it is susceptible to a number of illusions. These include:
1. The perceived size of an object may be influenced by its color.
2. The perceived hue of a color may be influenced by its saturation.
3. The perceived saturation of a color may be influenced by its hue.
4. The perceived depth of an object may be influenced by its color.
5. The perceived color of an object may be influenced by the color of surrounding objects.

Whenever possible, the conditions which give rise to these illusions should be avoided.

Color-size Effects. Some visual experiments have suggested that the color of an object can influence the perceived size of that object. Tedford, Berguist, and Flynn [77] surveyed studies of the effect of color on perceived size, noting that researchers differed in their conclusions about whether an effect existed as well as the relative ordering of color-size effects. They concluded that the disagreement could be attributed to lack of consistency of other stimulus characteristics such as saturation and brightness. They conducted their own experiments under precisely controlled conditions and found a significant color-size effect. Specifically, rectangles of the same size, saturation, and brightness appeared to have different sizes when colored red-purple, yellow-red, purple-blue, or green (in order of decreasing apparent size). At high saturations, this effect was statistically significant for all color pairs except yellow-red and purple-blue. At low saturations, only the difference between yellow-red and green rectangles was significant. In trials where hue was held constant and saturation varied, rectangles with higher saturations were consistently judged to be smaller than less saturated rectangles. Generalizing from the surveyed studies, they observed that warm colors (red, orange, yellow) appear larger than cool colors (blue, green).

Cleveland and McGill [83] investigated the implications of the color-size illusion for statistical maps. Subjects were shown a map of Nevada in which counties where colored red or green with the total area of red and green nearly equal. Subjects were asked to judge which color, if any, represented the larger land area. Each subject was shown ten maps. On the average, subjects judged that the red areas were larger more often than they judged the areas the same or the green areas larger. When the experiment was repeated using low-saturation tones of red and green (formed by adding yellow), no such bias was observed. Their results suggest that the color of a region influences the perceived size of the region, and the effect is strongest for very saturated colors.

2.5   Control level of detail

The amount of detail in a visualization should be appropriate to the form and content of informations displayed. Simply put : include enough detail, but not too much.

Detail in visualization can take many forms, including contour lines, surface detail, additional variables, and color scales with high frequency components.

Segmented vs. Holistic representation. Whether to represent a 2D or 3D field as discretized steps or as a continuous gradation depends on the necessity of displaying the segmented or holistic structure of the data. Contours or discretized color fields (2D datasets) and isosurfaces (3D datasets) present segmented structures; they utilize our perceptive ability to evaluate quantity at specific locations in 2D or 3D space. Continuous color fields (2D datasets) and voxel fields (3D datasets) present holistic structures; they utilize our perceptive ability to evaluate detailed form globally.

Figure 8 shows two displays of a pair of coexisting scalar datasets representing pollutant concentrations in 3D space. The isosurfaces in Figure 8a reveal the form of the SO_2 pollutant field at 4 and 8 ppb, and the SO_4 field at 4 ppb only. Obtaining precise information from the isosurfaces as to where the pollutants attain or exceed a given threshold (e.g. 4 ppb) is possible, but precisely determining the overall distribution of pollutants is not. By comparison, the voxel fields in Figure 8b reveal substantially detailed information of the pollutant structure in 3D space but do not clearly reveal any given quantity at a given location in space.

Recent perception research by Livingstone and Hubel [88] indicates that visual information is processed in at least three separate cognitive pathways in the human brain. One pathway, "blob-thin-stripe-V4,"processes perceived spatial distribution of colors. A second pathway, "parvo-interblob-pale-stripe-V4," evaluates high-resolution shape information. The third pathway, "magno-4B-thick-stripe," processes movement and stereoscopic depth. These three pathways are subsequently integrated so that one sees a unified environment in 3D space. This three-fold separation of visual information suggests that discrete representations of data (e.g. contours or isosurfaces) are likely processed primarily by a separate cognitive pathway than continuous data representations (e.g.. color or voxel fields): the shapes of contours and isosurfaces are processed by the shape-resolving "parvo" pathway, while the overall forms of continuous fields are processed by the "blob" pathway. Thus the choice of creating a discretized or continuous representation is tied to utilizing one of two distinct cognitive processes.

This distinction between two perceptual processes appears to be particularly striking when the datasets are animated as time sequences. When isosurface or contour representations are used and the datasets are animated, the viewer's attention seems to focus on the moving *edges* of the shapes rather than on the overall dynamics of the field. When continuous representations are utilized, the viewers attention seems to be more globally directed, making the perception of many different events simultaneously more possible than with discrete representations.

Color Scale Detail. Adding detail can sometimes make the fine structure of a value distribution more apparent. One way to do this is by using color scales with high frequency components. One type of high frequency color scale cycles rapidly and repeatedly through a sequence of colors. Such a scale makes small value differences more apparent because the colors representing them differ more than those from a lower frequency color scale. A repeating high frequency color scale, however, makes it impossible to look up the value represented by a color because a single color can represent several non-contiguous values. Adding non-repeating high frequency components such as contour lines or other sharp transitions can provide discrimination of small differences and metric lookup capabilities.

Figure 9a shows a visualization of suspended sediment concentrations using a spectrum color scale. It conveys a smoothly changing distribution where values drop off monotonically with distance from the two points of high value. Figure 9b represents the same data using a striped color scale. This color scale is composed of nine narrow bands of highly saturated color. Between each pair of bands is a section containing less saturated stripes of the two colors. In the image, the saturated bands give rough contours, the striped sections give an indication of the overall value distribution, and the stripes themselves show the fine structure of the value distribution.

Surface Detail. Adding surface detail to a height-mapped or iso-value surface can also facilitate the perception of fine surface structure. Figure 10 shows two height-mapped, pseudo-colored representations of ozone concentrations over the Mid-Atlantic states. The white lines below the ozone surface show state boundaries. In Figure 10a the transparent ozone surface allows the map to be seen, but small features of the surface are difficult to perceive. Figure 10b uses a texture map that modulates opacity to give the impression of a tangible surface while allowing the map to show through. Compare the lower right and upper left of the two images to see the differences in discernible detail of surface shape.

3 Multivariate Visualizations

Multivariate visualizations show two or more variables over a single spatial domain. The complexity of many scientific domains requires multiple variables to model phenomena or processes. For example, the complexity of environmental processes require simultaneous visualization of multiple quantities in order to represent the interactions among the various components. Multivariate visualizations in the environmental sciences can show joint distributions of two chemical species to compare patterns of pollution; airborne species in conjunction with weather conditions to see how weather affects pollutant transport; variables from two model domains (such as air and water) together in order to explore their interrelationships; species concentrations in conjunction with topography to see how terrain characteristics affect transport; or modeled data together with field samples of the same quantity for the purposes of model validation.

The basic challenges in the visualization of multivariate data are to:
1. Clearly show the spatial relationship of different variables.
2. Keep the different variable representations from interfering with one another. Specifically, the single-variable distributions should still be visible.
3. Facilitate the understanding of joint distributions.
4. Show as many variables as can be effectively displayed.

3.1 Show multiple surfaces

In some multivariate visualizations, the contributions of different variables naturally occupy different regions of space. Some examples are isosurfaces of different variables, isosurfaces at different value levels, situations where different variables are modelled or sampled at different locations, or height-surfaces of different variables. In such cases, the challenge is to display each variable in a way that does not obscure the others rather than to find a way to combine display parameters in the same space.

Most commonly, surfaces are made transparent so that other objects can be seen through them. A disadvantage of this technique is that discerning the shape of a transparent surface is more difficult than an opaque one. One reason for this is that obscuration cues are lost. Figure 11a shows the solvent-accessible surface of a molecule along with a ball and stick representation of the atoms and bonds. The solvent-accessible surface was made transparent to allow the ball and stick object to be seen. Figure 11b shows the same molecule with the surface textured with an opacity-modulated texture. Using this technique, the ball and stick object can be seen through the holes in the texture while the opaque sections provide surface shape cues. This technique also works for nested surfaces over an interior object (such as solvent-accessible surfaces using different radii), but the interior object is difficult to see clearly unless the representation is rotated interactively or in an animation.

3.2 Use orthogonal display parameters

Displaying two or more sets of data in a single visualization if often useful or necessary. Such an example was given in Figures 8a and 8b, which showed two pollutants coincident in 3D space. In such visualizations, a single consistent representation method (e.g. isosurfaces in Figure 8a) is most suitable for conveying the interaction between the multiple datasets. Often, however, the datasets will differ spatially or qualitatively, or will have little or no numerical correlation. It is necessary in these cases to represent the respective datasets with visual distinctness, or orthogonality to minimize visual confusion within the scene. In Figure 12, a visualization is constructed to convey three distinct numerical simulations simultaneously: pollutant concentration in 3D space, rainfall intensity on the 2D land projection, and pollutant deposition (also on the 2D land projection). The 3D pollutant data is mapped to an isosurface that propagates over the 2D land mass. Because the isosurface has distinct geometrical edges, representations for the 2D data which do not cause confusion with these geometric features are chosen. For the 2D deposition data, therefore, a continuous 2D color field is employed rather than contours because the gradual forms of the color field provide the needed visual distinctness to the well-defined form of the isosurface. Likewise, a segmented height field, composed of clear rectangular shafts, is chosen for the 2D rainfall data because its appearance provides significant orthogonality with respect to the other two representations.

In the perception model put forth by Livingstone and Hubel [88], the use of orthogonality is consistent with optimal utilization of the separate pathways of visual processing. According to this model, the distinct and sharp-edged shape of the isosurface in Figure 11 is processed largely by use of the shape-resolving parvo system. The form of the 2D pseudo-colored surface, with its gradual spatial changes in hue and value, is apprehended largely through use of the color-processing blob system, while the height/depth field of colorless shafts representing 2D rainfall is largely processed through the depth-processing magno system.

4   Conclusions

Attention to perceptual principles is essential to the construction of effective visualizations. These principles mandate the use of familiar paradigms, de-emphasis of uninteresting features, redundant mappings, appropriate level of detail, and orthogonal display parameters.

Acknowledgments

We would like to thank Dudley Bromley, Mark Bolstad, and Todd Plessel of Martin Marietta Technical Group at the EPA Scientific Visualization Center. We would also like to thank Terry Yoo for his technical wizardry and impressive patience.

Disclaimer

Although the research described in this article has been supported by the United States Environmental Protection Agency, it has not been subject to Agency review and, therefore, does not necessarily reflect the views of the Agency, and no official endorsement should be inferred. Mention of trade names or commercial products does not constitute endorsement or recommendation for use.

References

Cleveland, William S., and Robert McGill (1983), A Color-Caused Optical Illusion on a Statistical Graph, *The American Statistician*, vol. 37, no. 2, pp. 101-105.

Levkowitz, Haim (1988), Color in Computer Graphic Representation of Two-dimensional Parameter Distributions, PhD dissertation, University of Pennsylvania.

Livingstone, Margaret, and David Hubel (1988), Segregation of Form, Color, Movement, and Depth: Anatomy, Physiology, and Perception, *Science*, vol. 249, pp. 740-749.

Pizer, Stephen M., and John B. Zimmerman (1983), Color Display in Ultrasonography, *Ultrasound in Medicine and Biology*, vol. 9, no. 4, pp. 331-345.

Ramachandran, V. S. (1988), Perception of Shape from Shading. *Nature*, vol. 331, no. 6152, pp. 133-166.

Tedford, W. H. Jr, S. L. Berquist, and W. E. Flynn (1977), The Size-Color Illusion, *The Journal of General Psychology*, vol. 97, pp. 145-149.

Ware, Colin (1988), Color Sequences for Univariate Maps: Theory, Experiments, and Principles, *IEEE Computer Graphics and Applications*, Sept. , pp. 41-49.

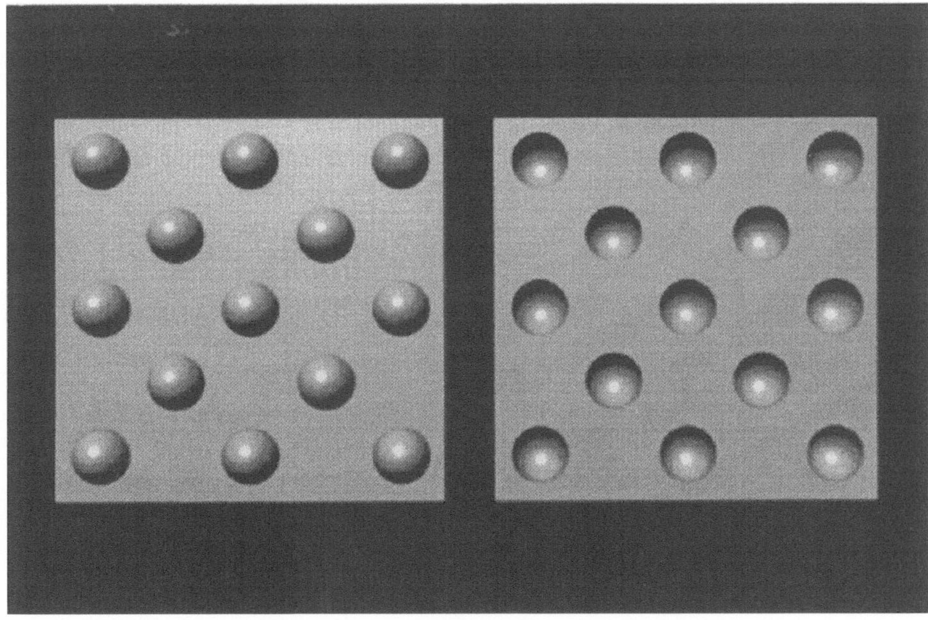

Figure 1. Two identical objects, illuminated from above (left) and from below (right).

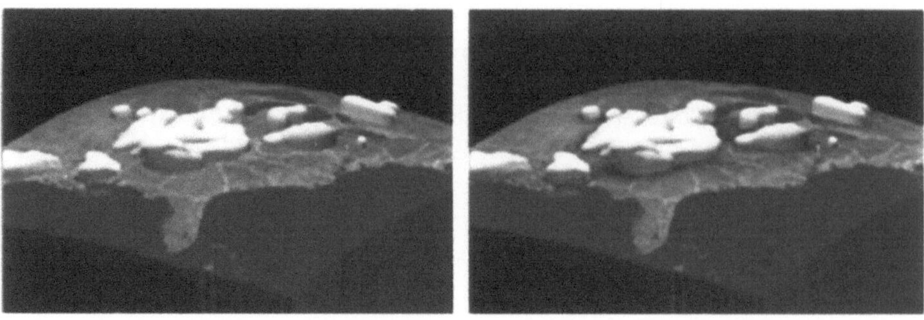

Figure 2. Visualization of an isosurface on a flat map surface: (a) without shadows and (b) with soft shadows projected onto the flat surface

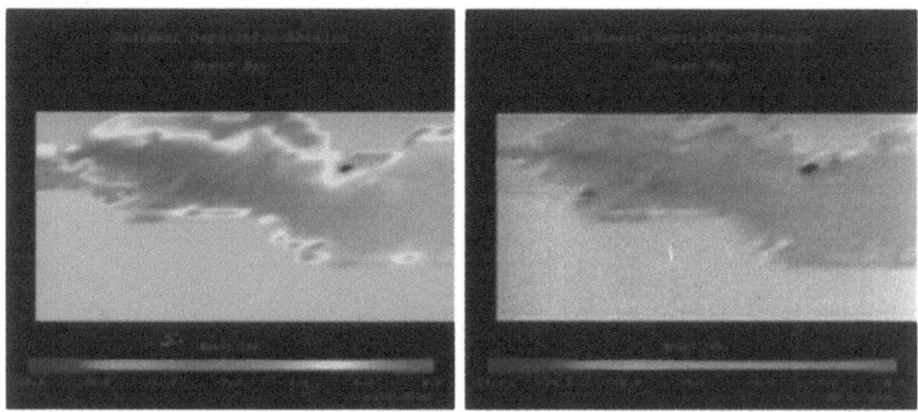

Figure 3. Sediment erosion/deposition data represented using a standard spectrum color scale (3a) and double-ended color scale (3b). Data courtesy of Joseph Gailani, Computer Sciences Corporation, EPA Large Lakes Research Station, Grosse Ile, MI

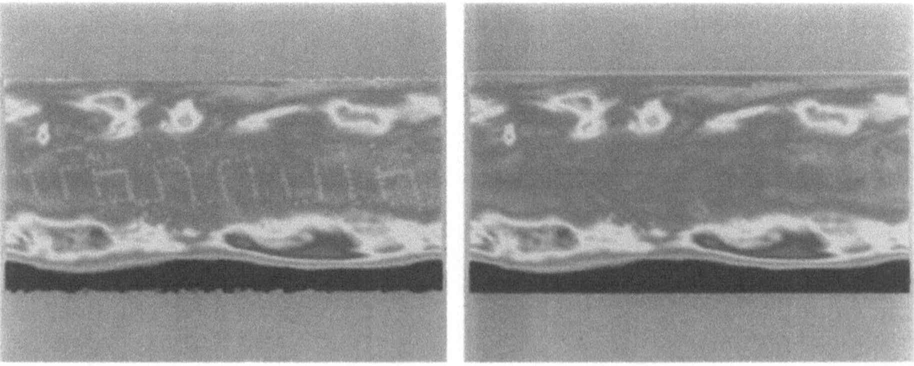

Figure 4. Southern hemisphere ozone concentrations with missing values blanked (4a) or interpolated (4b). Data courtesy of NASA Goddard. Images created by Mark Bolstad, Martin Marietta Services Group, EPA Scientific Visualization Center

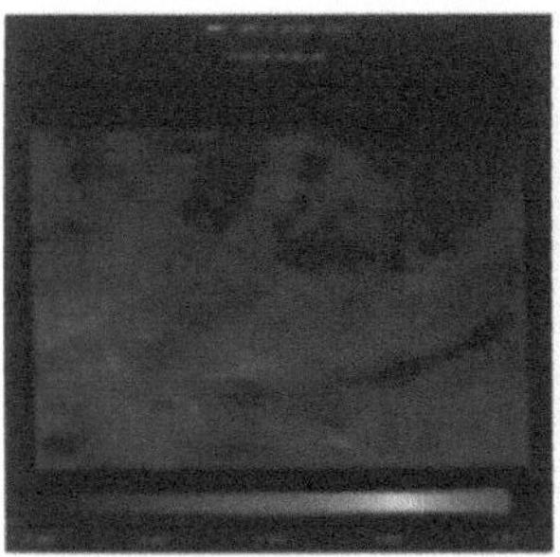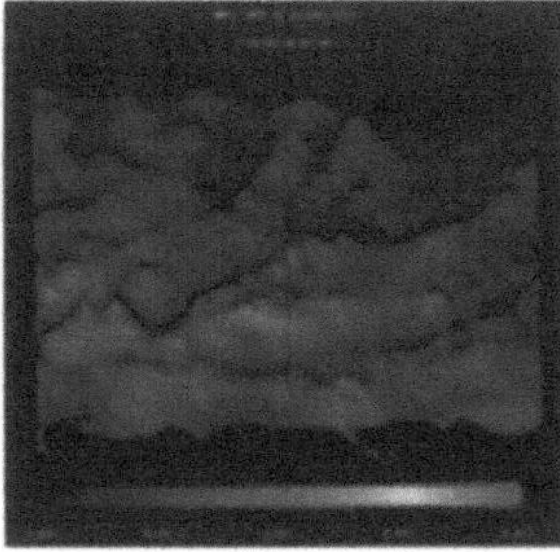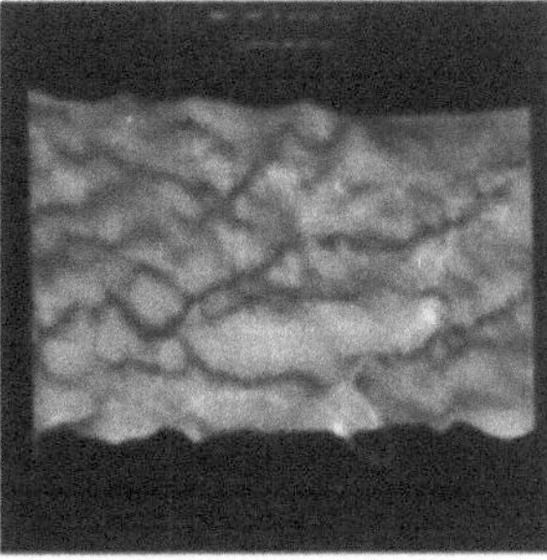

Figure 5. Concentrations of hydrogen peroxide represented using color (5a), height (5c), and a combination of both (5b). Data courtesy of U.S. EPA Atmospheric Research and Exposure Assessment Laboratory

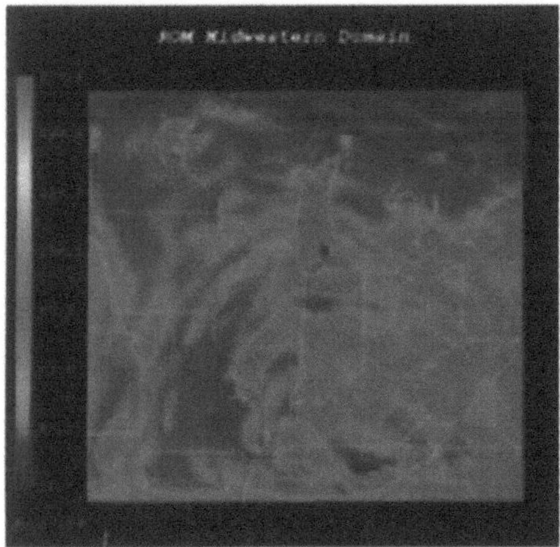

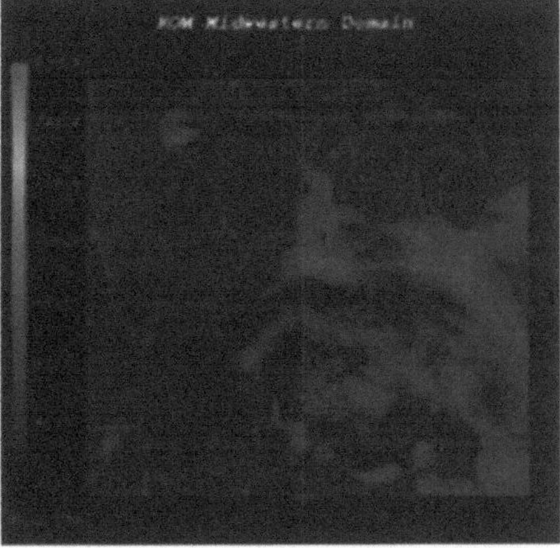

Figure 6. Ozone concentrations represented using standard spectrum (6a) and redundant rainbow (6b) color scales. Data courtesy of U.S. EPA Atmospheric Research and Exposure Assessment Laboratory

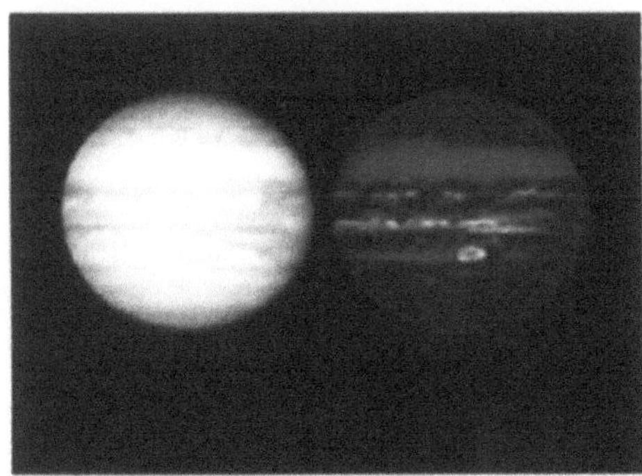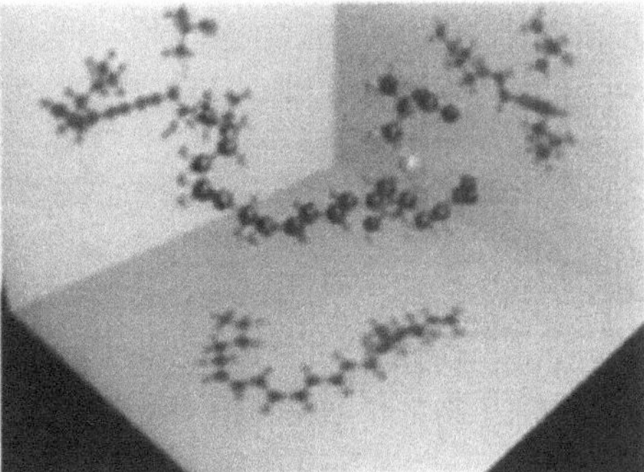

Figure 7. Two examples of explicit redundancy: a) Visualization of the planet Jupiter showing redundant representations (image courtesy of W. Lytle, Cornell Theory Center), b) display of the molecule LTE_4 showing redundant shadow projections on 3 surfaces (image courtesy of National Center for Supercomputing Applications)

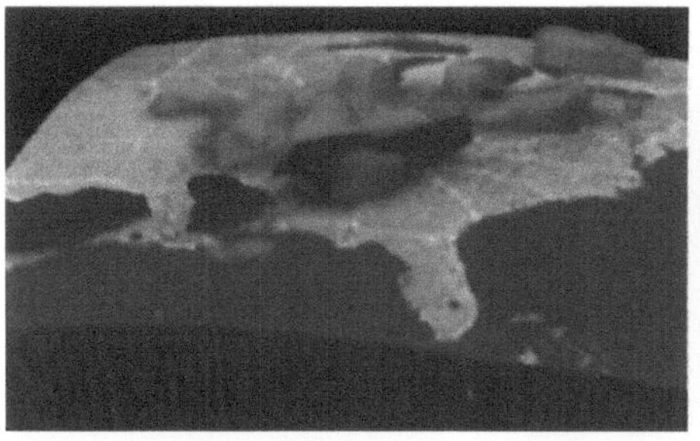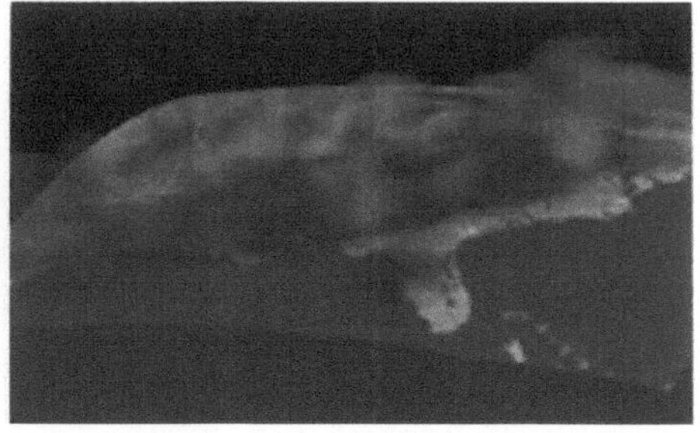

Figure 8. Comparison between 3D volumetric field visualizations. a) isosurface representation, b) voxel field representation

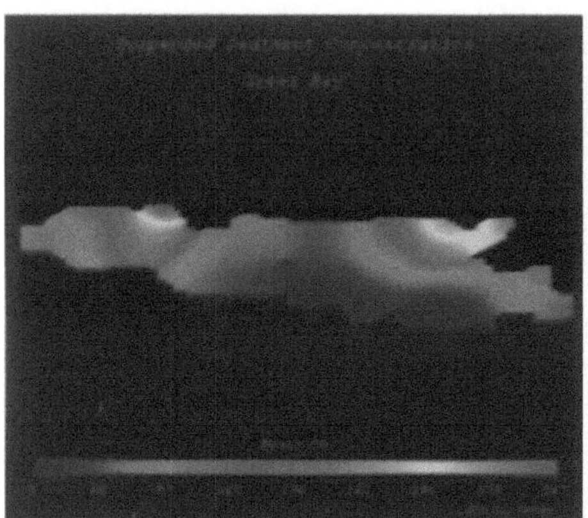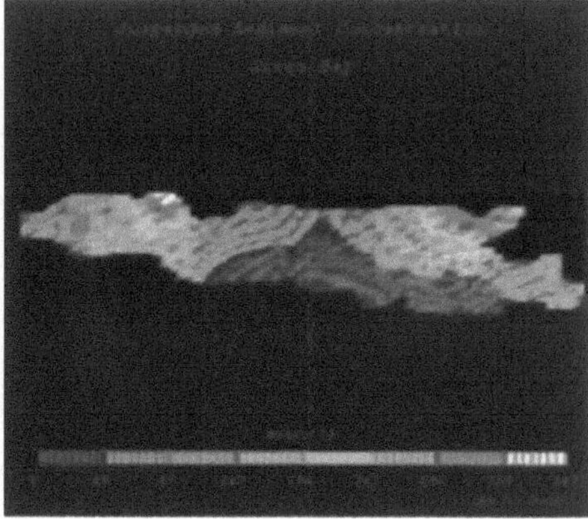

Figure 9. Suspended sediment concentrations represented using standard spectrum (9a) and striped (9b) color scales. Data courtesy of Joseph Gailani, Computer Sciences Corporation, EPA Large Lakes Research Station, Grosse Ile, MI

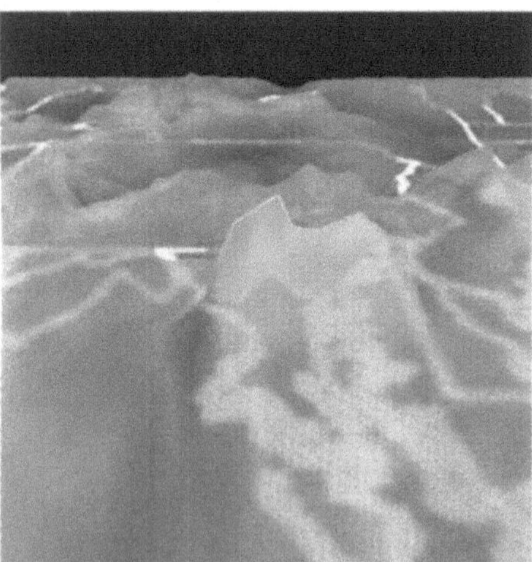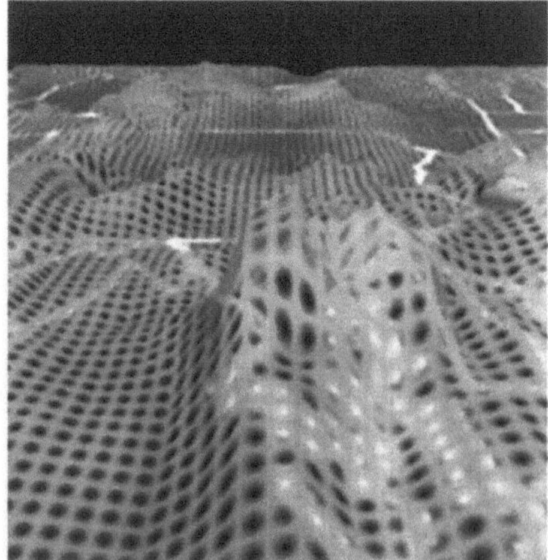

Figure 10. Ozone concentrations over Mid-Atlantic States represented using transparent surface (10a) and an opacity-modulated textured surface (10b). Data courtesy of U.S. EPA Atmospheric Research and Exposure Assessment Laboratory

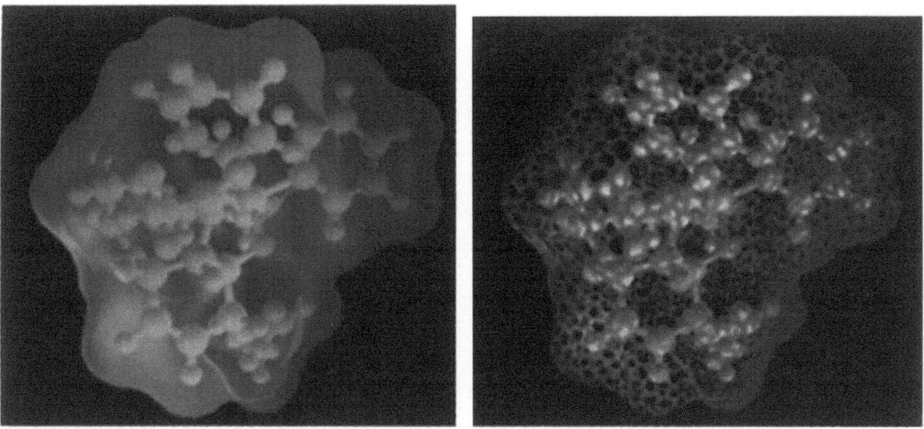

Figure 11. Molecule atoms, bonds, and solvent-accessible surface. The surface is either transparent (11a) or textured with an opacity-modulated texture (11b). Data courtesy of Mark Zottola, Duke University

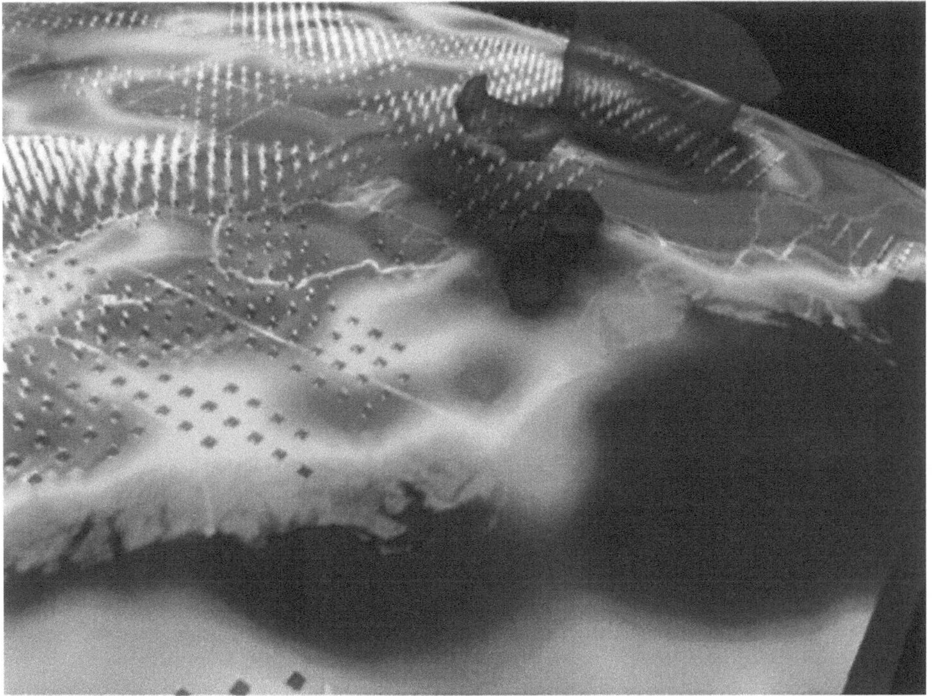

Figure 12. Use of orthogonality in layout. Simultaneous display of a 3D isosurface, 2D color field, and 2D height field

Interactivity and the Dimensionality of Data Displays

William L. Hibbard[1&2], Charles R. Dyer[2] and Brian E. Paul[1]

[1]Space Science and Engineering Center
[2]Computer Sciences Department
University of Wisconsin - Madison
Madison, WI 53706
whibbard@macc.wisc.edu

Abstract. Using mathematical models of data and displays, we illustrate the importance of distinguishing between independent and dependent variables when counting the dimensions of data sets and displays. The number of independent variables occurring as dimensions of a display model is the most important factor determining its information carrying capacity. Independent variables in a display model also require interactive techniques for their implementation, as illustrated by our VIS-AD system and Beshers' and Feiner's *worlds within worlds* technique. Thus interactivity is critical for visually communicating large amounts of information, and the perceptual properties of interaction techniques are an important topic for visualization research.

1 Introduction

There is a constant need to increase the capacity of visualization techniques to handle large and complex data sets. This need is an important motive for this workshop on perceptual issues in visualization. We will use a mathematical model of the visualization process to analyze the information carrying capacity of displays, and to understand the impact of various display techniques on that capacity. In particular, we will discuss the way that data set dimensions are counted, and the importance of interactivity for displaying data sets of high dimensionality. We will illustrate these ideas using our VIS-AD system, and using Beshers' and Feiner's *worlds within worlds* technique.

2 A Formal Model of the Display Process

We model a data display process as a function $D:U \rightarrow V$, where U is a set of data objects, called a data model, and V is a set of displays, called a display model. Data models define and organize sets of data objects, and display models define and organize sets of displays. Computer generated displays exist as data objects inside of computers, which suggests that we can use the same formalism for data models and for display models. Data models need to specify the types of primitive values occurring in data objects, and how those values are aggregated into data objects.

Thus we define a set S of scalar data types, whose values are the primitive values of our data model, and a larger set T of complex data types, whose values are aggregates of primitive values.

1. Define a set S of finite primitive sets, where S is a subset of T. We call types in S scalar types. Each scalar type specifies a set of primitive values. For example, a scalar type may be a set of 32-bit integers, it may be a set of 8-bit gray scale intensities, or it may be a set of 1024 temperatures indexed by a 10-bit satellite radiance. A scalar set of primitive values may include a distinguished "missing" value, indicating the absence of information. This can be useful for satellite images and other data generated by fallible remote sensing instruments.

2. Given sets A_1, A_2, ..., A_n in T, the cross product $(A_1 \times A_2 \times ... \times A_n)$ is in T. We call $(A_1 \times A_2 \times ... \times A_n)$ a tuple type, with element types A_1, A_2, ... and A_n. For example, if CR, CG and CB are sets of red, green and blue intensities, then the tuple type $(CR \times CG \times CB)$ is a pixel color.

3. Given sets A in S and B in T, the function space $(A \rightarrow B)$ is in T. We call $(A \rightarrow B)$ an array type, with domain type A and range type B. It is the set of functions from A to B. For example, if TEMP is a set of temperatures and TIME is a set of times, then the array type $(TIME \rightarrow TEMP)$ is a set of time series of temperatures.

Array types define functional relations from their domain types to their range types. An array data object contains one object of the range type for each value of the domain type, and thus defines a function from the domain's scalar value set to the set of range data objects. The array's domain type is the independent variable of this functional relation. In a complex type, scalar types occurring as array domains are independent variables, and all other scalar type occurrences are dependent variables.

In [3] we study general properties of display functions of the form $D:U \rightarrow V$, in terms of a data model U that includes all types in T. Here we focus on the specific issues of defining the information carrying capacity of display models and defining data set dimensionality. For this focus we can study the simple situation where U and V are specific types in T.

We define, as a condition on D, that users should be able to distinguish different data objects from their displays. This just says that D maps different data objects to different displays, or, in mathematical terminology, that D is injective. If $|U|$ and $|V|$ denote the cardinalities of U and V, then this condition requires that $|U| < |V|$. Thus, we can consider the cardinality $|V|$ as a kind of information carrying capacity of a display model.

The cardinalities of types in T can be calculated using the following two equations for the cardinalities of tuple and array types:

$$(1) \qquad |(A_1 \times A_2 \times ... \times A_n)| = |A_1| * |A_2| * ... * |A_n|$$

$$(2) \qquad |(A \rightarrow B)| = |B| \uparrow |A|$$

Note that the cardinality of the array domain A has a much larger effect than the array range B on the cardinality of the array type $(A \rightarrow B)$.

3 The VIS-AD Display Model

We can define a simple example of a display model $V_{2\text{-}D}$ in T. Let XH, XV, CR, CG and CB be sets in S, where XH and XV are sets of horizontal and vertical screen coordinates, and CR, CG and CB are sets of red, green and blue intensities. Then we can define a display model whose members are two-dimensional arrays of colors, as follows:

$$(3) \qquad V_{2\text{-}D} = (XH \rightarrow (XV \rightarrow (CR \times CG \times CB)))$$

The displays in $V_{2\text{-}D}$ are static 2-D images. In order to elaborate this display model to include three dimensions, animation and user interaction, we define the following scalar types:

(4)	X, Y, Z	sets of values for three voxel coordinates
(5)	CR, CG, CB	as before, sets of voxel color intensities
(6)	$C_1, C_2, ..., C_m$	sets of values depicted by iso-level contours
(7)	AN	a set of animation step numbers
(8)	$S_1, S_2, ..., S_n$	sets of values used as abstract indices of display contents

Then we define the display model $V_{VIS\text{-}AD}$ by:

$$(10) \qquad VOXEL = (CR \times CG \times CB \times C_1 \times C_2 \times ... \times C_m)$$
$$(11) \qquad VOLUME = (X \rightarrow (Y \rightarrow (Z \rightarrow VOXEL)))$$
$$(12) \qquad SEQUENCE = (AN \rightarrow VOLUME)$$
$$(13) \qquad V_{VIS\text{-}AD} = (S_1 \rightarrow (S_2 \rightarrow ... (S_n \rightarrow SEQUENCE)...))$$

A data object in VOLUME is a 3-D volume of voxels that is projected onto a 2-D display screen. A data object in SEQUENCE is a sequence of voxel volumes that can be displayed in quick succession to create an animated display. A data object in $V_{VIS\text{-}AD}$ is a set of nested arrays that contains many 3-D animation sequences. By selecting an index value for each nested array (i.e., a value from each S_i) users select one sub-object of type SEQUENCE (an animation sequence) from an object of type $V_{VIS\text{-}AD}$. The data objects in $V_{VIS\text{-}AD}$ are very large, so we do not assume that data objects from $V_{VIS\text{-}AD}$ ever exist all at once. Rather, $V_{VIS\text{-}AD}$ is an abstract model of display behavior, and an actual implementation only needs to generate a part of a data object in $V_{VIS\text{-}AD}$ at any one time [2].

By applying equations (1) and (2) we can calculate $|V_{VIS\text{-}AD}|$ as:

$$(14) \qquad |V_{VIS\text{-}AD}| = (|CR| * |CG| * |CB| * |C_1| * |C_2| * ... * |C_m|) \uparrow$$
$$(|X| * |Y| * |Z| * |AN| * |S_1| * |S_2| * ... * |S_n|)$$

The array domains X, Y, Z, AN, S_1, S_2, ..., S_n all appear in the exponent of equation (14) and thus have the largest impact on $|V_{VIS-AD}|$. They are the independent variables of the display model. The color intensities CR, CG and CB, and the contour variables C_1, C_2, ... and C_m, are dependent variables, and have a smaller impact on $|V_{VIS-AD}|$.

The definition of V_{VIS-AD} models animated three dimensional displays, and also models certain forms of user interaction. For example, users control the projection of the voxel volume onto a two-dimensional screen. That is, users control the projection of X, Y and Z onto screen coordinates XV and XH. Users control animation by controlling the sequencing of indices in AN. The scalars S_1 through S_n are called "selectors" and provide an abstract model of user control over display contents. By selecting values for S_1 through S_n, a user selects a particular member of SEQUENCE, and thus controls the contents of the display screen. Thus the display model V_{VIS-AD} models interactive rotation, animation control, and selection between alternate display contents.

4 Defining the Dimensionality of Data and Display Models

In either a data model or a display model, it is important to make a clear distinction between independent variables, which are array domains and appear in the exponent of data set cardinality, and dependent variables, which are array ranges or tuple elements. In equation (14) for the cardinality $|V_{VIS-AD}|$, the independent variables have a much larger impact on $|V_{VIS-AD}|$ than the dependent variables do. From the point of view of data set size and the information carrying capacity of displays, it is inappropriate to mix independent and dependent variables in counting data set dimensionality.

In fact, data dimensionality should only count nested independent variables, since it is the product of the sizes of nested independent variables that dominate the size of the exponent in the expression for data type cardinality. For example, consider the data type:

$$(15) \qquad G = (A \rightarrow ((B \rightarrow (C \times D)) \times (E \rightarrow F)))$$

Applying equations (1) and (2), the cardinality of G is:

$$(16) \qquad |G| = (((|C| * |D|) \uparrow (|B| * |A|)) * (|F| \uparrow (|E| * |A|))$$

This is a product of two terms. The exponent in the first term is $|B| * |A|$ and the exponent in the second term is $|E| * |A|$. Since each of these exponents is the product of two primitive set sizes, we should only count the dimensionality of G as two. Although A, B and E are all array domains and thus independent variables, B and E are not nested. Of course, there are other factors in evaluating the complexity of G. For example, the type G is certainly more complex than the two-dimensional data type H = (A → (B → C)). However, from the point of view of cardinality, the data types G and H are roughly similar.

Much work on perception focuses on attributes of voxels, such as color, texture and shape. While these attributes, intelligently employed, are very useful for addressing the logical complexity of a data model, they are dependent variables in a display model. Thus, they have limited usefulness for addressing the cardinality problem created by data models with high dimensionality.

In order to address the problem of visualizing data models with high dimensionality, it is more effective to add nested independent variables to a display model than it is to add perceptual attributes of pixels. The scalars X, Y, Z, AN, and S_1 through S_n are nested independent variables in the definition of V. This is in fact the display model used for the VIS-AD system [2]. Since people only directly sense three spatial dimensions plus one time dimension, display models with more than four dimensions must depend on interaction to allow users to explore their higher dimensionality. As we described earlier, interaction techniques for the VIS-AD display model include rotation and zoom of the 3-D voxel volume, animation control, and selecting values for S_1 through S_n.

5 The Worlds Within Worlds Display Model

Beshers and Feiner [1] have defined a display model that they called *worlds within worlds* and that embedded small 3-D coordinate systems (complete with little sets of three axes) inside larger 3-D coordinate systems. At the innermost level of embedded coordinate systems, data are generally depicted as surfaces, with Z as a function of X and Y. Several such surface plots may be depicted within a containing coordinate system, with the origin of each inner system set at the appropriate coordinates within the containing system. Users can drag the inner systems around within the containing systems, in order to explore the higher-dimensional spaces. Using the notation of our set of types T, Beshers' and Feiner's display model can be roughly defined by:

(17) $V_1 = (X_1 → (Y_1 → Z_1))$
(18) $V_2 = (X_2 → (Y_2 → (Z_2 → V_1)))$

    . . .

(19) $V_{WORLDS} = V_n = (X_n → (Y_n → (Z_n → V_{n-1})))$

This display model has 3 * n - 1 independent variables, and Z_1 is its only dependent variable.

6 The Importance of Interactivity

The data type V_{WORLDS} defined for the *worlds within worlds* display model, and the data type V_{VIS-AD} defined for the VIS-AD display model, share the property that the number of nested independent variables can vary to accommodate the dimensionality of the data set being displayed. That is, the display model

$V_{VIS\text{-}AD}$ includes a variable number of nested selectors S_i, and the display model V_{WORLDS} includes a variable number of nested coordinates X_i, Y_i and Z_i. Increasing the number of these variables increases the information carrying capacities $|V_{VIS\text{-}AD}|$ and $|V_{WORLDS}|$.

Since perceptions are limited to three spatial dimensions plus one time dimension, display models must rely on interactive techniques to depict more than four nested independent variables. In the VIS-AD display model the user interactively controls ranges for the selectors S_i. In the *worlds within worlds* display model the user interactively moves inner coordinate systems to select values in containing coordinate systems. In both of these systems the user interactively selects values for independent variables of the display model. These selected values are used to extract display objects of smaller dimensionality from the larger display model, according to functional relations defined in the display model. This dimensionality reduction results in a display object that is appropriate for direct animated 3-D display.

7 The Mapping from Data Model to Display Model

In general, there is a choice of many different functions $D:U \to V$ mapping from a data model to a display model. The VIS-AD system gives the user interactive control over this choice. This is in addition to the interactive techniques built into the VIS-AD display model that are described in Section 3.

The VIS-AD system integrates a computational model, a data model and a display model, in order to help scientists conduct visual experiments with their algorithms. The computational model is a programming language similar to C. The data model includes data objects of all types in T (as defined in Section 2), and these are just the data objects of the VIS-AD programming language. The type $V_{VIS\text{-}AD}$ is the VIS-AD display model. Users experiment with their algorithms by making changes to their algorithms and interactively executing them. There are also tools in VIS-AD to let users alter parameters of algorithms as they run (i.e., interactively steer computations). Because of the level of abstraction of the VIS-AD data model and display model, any combination of data objects can be displayed during algorithm execution. This provides a powerful way for users to visualize the behavior of their algorithms.

An important feature of VIS-AD is the fact that program data objects can be displayed without embedding calls to display functions in users' programs. Thus user's are free to concentrate on the scientific content of their algorithms without the need to intermix display commands. The key to this capability is the level of abstraction of the display process as a function $D:U \to V_{VIS\text{-}AD}$. Users specify D in terms of the set S of scalar types, and this defines $D:U \to V_{VIS\text{-}AD}$ for all types U that can be derived from S. The left hand column of Figure 1 illustrates two data types defined for an algorithm for discriminating clouds in multi-spectral satellite images. The right hand column of Figure 1 is a diagram of the VIS-AD display model. Users specify display functions D in terms of a set of mappings from the set

S of scalar types to the set of display scalar types defined in (4) through (8), as illustrated in the center column of Figure 1.

The key idea of the VIS-AD system is that the display functions $D{:}U \rightarrow V_{VIS\text{-}AD}$ can be controlled by defining mappings between scalars. Of course, there are other approaches to providing interactive user control over the display function D. The data flow systems let users control the way that data are displayed by editing a data flow network. However, this is essentially a programming interface, and mixes display logic with data analysis logic. The scalar mappings of VIS-AD separate display logic (which should not require a programming interface) from analysis logic (which does require a programming interface).

8 Conclusions

Designing interaction techniques that effectively present large numbers of dimensions, defined as nested independent variables in a display model, is a challenging perceptual problem, but is also the key problem for visualizing data sets with high dimensionality.

Because only a finite amount of information can be displayed at any instant, display models need to include interactive techniques in order to depict large data sets. The information carrying capacity of display models can be increased by increasing the number of nested independent variables they include. But because our perceptual capabilities are limited to three space dimensions plus one time dimension, display models of high dimensionality require interactive techniques. Furthermore, interaction techniques can include user control over the functional form of the mapping from data to displays.

References

1.	Beshers, C. and S. Feiner, 1992; Automated design of virtual worlds for visualizing multivariate relations. Proc. Visualization '92, IEEE, 283-290.

2.	Hibbard, W., C. Dyer and B. Paul, 1992; Display of scientific data structures for algorithm visualization. Proc. Visualization '92, Boston, IEEE, 139-146.

3.	Hibbard, W., C. Dyer and B. Paul, 1994; A Lattice Model for Data Display. In Preparation.

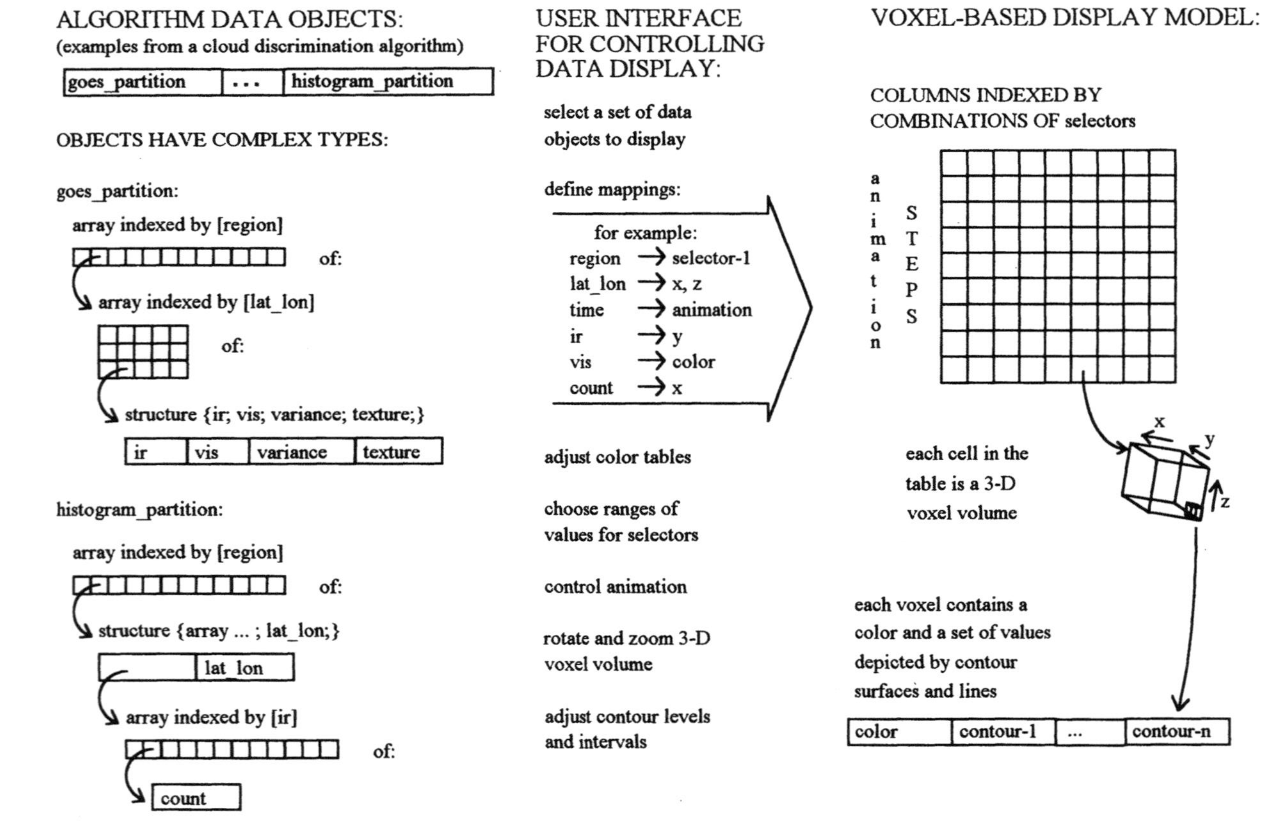

FIGURE 1. Examples data types that users can define in the VIS-AD system, a diagram of VIS-AD's voxel-based display model, and the way that users control how data are mapped into the display model. No graphics logic is required from the user, other than to define mappings of scalar types into VIS-AD's display model.

Towards Perceptual Control of Markov Random Field Textures

Rongxin Li and Philip K. Robertson

CSIRO Division of Information Technology
P.O.Box 664, Canberra, ACT 2601, Australia
ron.li@csis.dit.csiro.au phil.robertson@csis.dit.csiro.au

Abstract. In this paper we propose a method of establishing a perceptually meaningful Euclidean space for textures generated as samples of first-order Markov Random Fields (MRF). We show how, under a definition of texture difference, within a specified neighbourhood of zero the first-order MRF parameters can be considered as orthogonal and approximately define textures in a uniform manner. We suggest that the Euclidean space established with this texture distance metric may be considered a perceptual texture space that can be used for visualisation purposes. The definition of texture difference is based on texture classification results that are consistent with human visual discrimination when experimented with a set of test images. Demonstrations of using our texture space as a preliminary visualisation tool are presented.

1 Introduction

Texture provides rich information about visual scenes. This suggests that texture could also be used effectively to convey information in data and process visualisation. Experiments with various texture representations in visualisation have been performed [1, 2]. For texture to be an effective and reliable information carrier, however, systematic control of texture generation to meet perceptual assessment criteria is required.

In this work we investigate the possibility of developing a perceptual texture space to allow the mapping of data into defined perceptual parameters of texture in a uniform way. Markov Random Fields (MRFs) provide a powerful computational tool for generating textures in a systematic manner, enable the generating (sampling) process to be computationally local and parallel, and have substantial applications in image modelling [3, 4]. In this paper we investigate systematically how variations of MRF parameters result in changes in the perception of the realization.

The Markov model of a random field $X = \{X_{i,j}\}$ defined on a finite lattice $L = \{(i, j)\}$ with each $X_{i,j}$ being assigned a value from a finite set A prescribes three criteria: (1) that the assignment of any value in A to any $X_{i,j}$ must be possible (positivity); (2) that the probability of such an assignment in the presence of the assigned values at all other sites only depends on the assignments to the neighbours of $X_{i,j}$ (Markovianity); (3) that this dependence relationship is invariant over L (Homogeneity). Formal mathematical definitions are given by Rozanov [5] and by Dubes and Jain [6].

We take the approach of developing a relationship between MRF parameters and psychophysically validated computational measures of texture perception. We choose image classification based on local neighbourhood value dependency matrices. We investigate classification measures, explained using second order statistics, that have been used for successful texture discrimination, and then relate these to the second order statistics of the MRF. Assumptions and limitations are summarised, and some simple demonstrations are provided to show the current status of the work.

2 An image classification scheme

The use of certain feature matrices of images, such as the neighbouring gray-level dependence matrix (NGLDM) and the spatial gray-level dependence matrix (SGLDM), has resulted in considerable success in classification of textured images [7]. For example, the NGLDM, when used in conjunction with a maximum likelihood texture classifier, has been shown to be able to classify sample sets of distinct synthetic and natural textures with 100% accuracy. We therefore to use those matrices as perceptually valid representation of textures.

Whereas for image classification tasks one needs rotation-invariant measures, this is not the case for a visualization task. In fact, while a rotated image may belong to the same class as the original one, the two images may have different meanings and are often distinguishable by humans. We therefore need angularly dependent measures. The SGLDM has an angularly dependent version. As the NGLDM was designed to be rotationally invariant, we need to modify it to induce angular dependence.

The SGLDM that shows gray-level co-occurrence of sites aligned in direction α with distance d between them is defined such that the element in the mth row and nth column is given by

$$SGLDM\,(m, n\mid d, \alpha) \;=\; Card\,\{\,(i, j)\,,\,(k, l) \in L\mid \rho\,(\,(i, j),\,(k, l)\,) \;=\; d,$$
$$angle\,(\,(i, j),\,(k, l)\,) \;=\; \alpha,\, H\,(i, j) \;=\; \Phi_m,\, H\,(k, l) \;=\; \Phi_n\,\}$$

where Card(A) is the cardinality of set A (i.e. the total number of elements in A), L is the lattice underlying a digitized image, $H(i, j)$ and $H(k, l)$ are the gray levels at (i, j) and (k, l) respectively, Φ_m and Φ_n represent the mth and nth gray levels, $\rho\,((i, j), (k, l))$ is the distance between points (i, j) and (k, l). Haralick et. al., who first introduced the SGLDM, use the square distance metric (i.e. $max\{li\text{-}kl, lj\text{-}ll\}$)[8].

In the original NGLDM, for given parameters d (the neighbourhood size) and a (the range of dependency - the co-occurrence of two gray levels is counted if the difference between them falls into this range), the element in the mth row and nth column is given by

$$NGLDM\ (m, n|\ d, a)\ =\ Card\ \{\ (i, j)\ \in\ L|\ H\ (i, j)\ =\ \Phi_m,$$
$$Card\ \{\ (k, l)\ \in\ L - \{\ (i, j)\ \}\ |\ \rho\ (\ (i, j),\ (k, l)\ )\ \leq d,$$
$$|H\ (i, j)\ -\ H\ (l, k)|\ =\ a\ \}\ =\ n\ \}\ .$$

To introduce angular dependence, we re-define the NGLDM elements with an additional directional parameter α:

$$NGLDM\ (m, n|\ d, \alpha, a)\ =\ Card\ \{\ (i, j)\ \in\ L|\ H\ (i, j)\ =\ \Phi_m,$$
$$Card\ \{\quad (k, l)\ \in\ L - \{\ (i, j)\ \}\ |\ \rho\ (\ (i, j),\ (k, l)\ )\ \leq d,$$
$$angle\ (\ (i, j),\ (k, l)\ )\ =\ \alpha,\ |H\ (i, j)\ -\ H\ (l, k)|\ =\ a\ \}\ =\ n\ \}$$

In Berry & Goutsias' approach [7] to texture classification problems, a "representative" matrix was estimated from a few training samples from each class, and the following log-likelihood classifier was used:

$$L_k\ (Q_m) = \Sigma\ \left(Q_m\ (i, j)\ \ln\ \{\ Q_k\ (i, j)\ /\ [\ \Sigma\ Q_k(i, j)\ ]\ \} \right),$$

where Q_m is the feature matrix derived from the image to be classified, Q_k is estimated from a set of training samples from class k and is considered the representative matrix of class k, $Q_m(i, j)$ and $Q_k(i, j)$ are the corresponding matrix elements of Q_m and Q_k respectively, and Σ represents summation over all elements in the matrices. The largest $L_k(Q_m)$ puts the image into class k.

3 The second-order statistics in samples of MRFs

$L_k(Q_m)$ reflects the degree of perceptual similarity between textures. We hypothesise that we can determine perceptual differences between realizations of different MRFs if we can predict corresponding neighbouring gray level dependence matrices from the MRF parameters. The key problem in this prediction is to determine the dipole statistics (i.e. the second-order statistics) between two nearest points.

The MRF-GRF (i.e. the Gibbs Random Field) equivalence ensures that the global distribution of an MRF is

$$p(X) = 1/Z\ \exp\ \{ -a\ \Sigma\ X_{i,j}\ -\ b_1\ \Sigma\ X_{i,j}\ X_{i-1,j}\ -b_2\ \Sigma\ X_{i,j}\ X_{i,j-1} \}$$

where a, b_1 and b_2 are the first order MRF parameters, and

$$Z = \Sigma \exp \{-a \Sigma X_{i,j} - b_1 \Sigma X_{i,j} X_{i-1,j} - b_2 \Sigma X_{i,j} X_{i,j-1}\}$$

is the partition function, which is the sum of the numerator exponents over all possible images. We note that the expectation of the sum of the products of nearest horizontal neighbours can be calculated from the partial derivative of the partition function with respect to the corresponding first-order MRF parameter:

$$E \{\Sigma X_{i,j} X_{i-1,j}\} = - \partial \{\log[Z(a, b_1, b_2)]\} / \partial b_1 .$$

The difficulty is that the partition function Z cannot be computed. Fortunately, in the literature of statistical physics, Onsager [9], Newell and Montroll [10], and Bartlett [11] have by different means found the solution for the case of the first-order MRF, given a=0. Assuming an $N = M_1 \times M_2$ rectangular lattice, on which the value at each point can be either 1 or -1, it has been proved that

$$-\log Z = N\log 2 + \sum_{k=1}^{\frac{M_1}{2}} \sum_{l=1}^{M_2} \log \left(\cosh 2b_1 \cosh 2b_2 - \sinh 2b_1 \cos \frac{4\pi k}{M_1} \right.$$
$$\left. - \sinh 2b_2 \cos \frac{2\pi l}{M_2} \right).$$

Therefore if we use q to denote $E \{(\Sigma X_{i,j} X_{i-1,j})/N\}$, we have

$$q = \frac{1}{N} \sum_{k=1}^{\frac{M_1}{2}} \sum_{l=1}^{M_2} \frac{2\sinh 2b_1 \cosh 2b_2 - 2\cosh 2b_1 \cos \frac{4\pi k}{M_1}}{\cosh 2b_1 \cosh 2b_2 - \sinh 2b_1 \cos \frac{4\pi k}{M_2} - \sinh 2b_2 \cos \frac{2\pi l}{M_2}} .$$

When the lattice is large enough,

$$q = \frac{1}{\pi^2} \int_0^\pi \int_0^\pi \frac{2\sinh 2b_1 \cosh 2b_2 - 2\cosh 2b_1 \cos w_1}{\cosh 2b_1 \cosh 2b_2 - \sinh 2b_1 \cos w_1 - \sinh 2b_2 \cos w_2} dw_1 dw_2 .$$

We perform the above integration numerically because of the difficulty in finding the analytical solution. Figure 1 shows q plotted against b_1 and b_2. We observe that (1) q is mainly affected by b_1, and (2) q is roughly a sigmoid function of b_1, consisting of a nearly linear region in the middle and two saturated regions (in fact, when b_2=0 q is one of the typical sigmoid functions - tanh b_1).

For $b_1 \in$ [-0.5, 0.5], it can be shown that

$$q \approx 1.1838\, b_1$$

with an average square-root error of 0.068. Relatively large errors only occur when b_2=0.

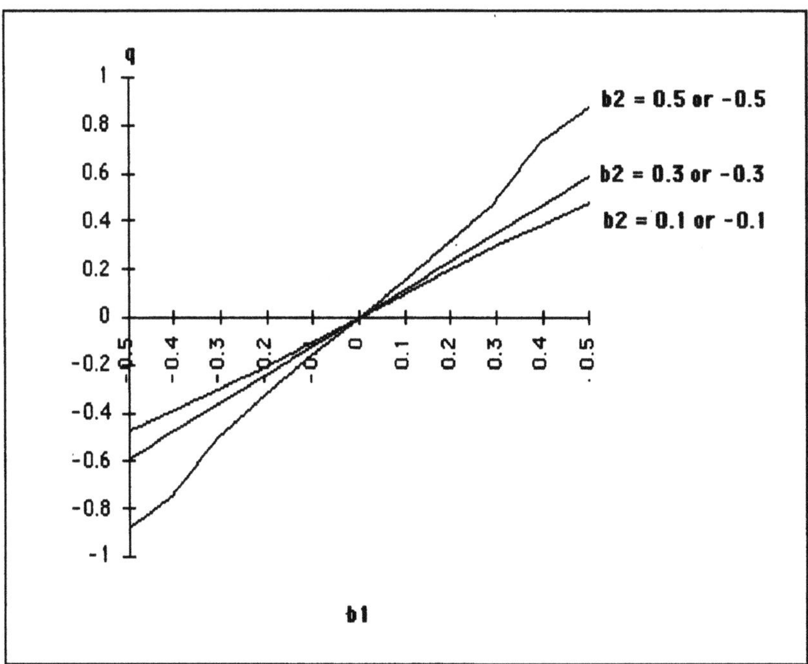

Fig. 1. When b_1 and b_2 are within a restricted neighbourhood of zero (e.g. [-0.3, 0.3]), q is approximately a linear function of b_1 and almost unaffected by the variation of b_2.

Letting P_1 denote the probabilities

$$P_1 = p\,(X_{i,j} = X_{i-1,j} = 1) = p\,(X_{i,j} = X_{i-1,j} = -1),$$

since

$$E\,\{(\Sigma\, X_{i,j}\, X_{i-1,j})/N\} = 2\,P_1 - (1 - 2P_1),$$

we have

$$P_1 = (q+1)/4 \approx 0.25 + 0.296\, b_1.$$

Similarly, if we let P_2 denote $p(X_{i,j} = X_{i,j-1} = 1)$ or $p(X_{i,j} = X_{i,j-1} = -1)$, then

$$P_2 \approx 0.25 + 0.296\, b_2.$$

4 The expected SGLDM and NGLDM of an MRF texture

Much evidence supports the view that visual discrimination is based upon local features or local differences in the second-order spatial averages [12, 13, 14]. Therefore we use the matrices with d=1. The SGLDM (d=1, $\alpha=180^\circ$) in this case, using a normalized version of SGLDM in which the elements are all divided by N, is:

$$\begin{bmatrix} (0.25 + 0.296 b_1) & (0.25 - 0.296 b_1) \\ (0.25 - 0.296 b_1) & (0.25 + 0.296 b_1) \end{bmatrix}.$$

The SGLDM (d=1, $\alpha=90^\circ$) is

$$\begin{bmatrix} (0.25 + 0.296 b_2) & (0.25 - 0.296 b_2) \\ (0.25 - 0.296 b_2) & (0.25 + 0.296 b_2) \end{bmatrix}.$$

The normalized NGLDM(d=1, $\alpha=180^\circ$, a=0) is

	n=0	n=1	n=2
m=-1	$(1 - 2P_1)^2$	$4P_1(1 - 2P_1)$	$4P_1^2$
m=1	$(1 - 2P_1)^2$	$4P_1(1 - 2P_1)$	$4P_1^2$

The normalized NGLDM(d=1, $\alpha=90^\circ$, a=0) is

$$\begin{bmatrix} 4P_2^2 & [4P_2(1 - 2P_2)] & (1 - 2P_2)^2 \\ 4P_2^2 & [4P_2(1 - 2P_2)] & (1 - 2P_2)^2 \end{bmatrix}.$$

5 A Euclidean space for the first-order MRF textures

Given the success of the classification scheme employing the NGLDM or the SGLDM, and the log-likelihood classifier, we suggest that it is perceptually meaningful to define a distance $D(m,n)$ between textures m and n as follows:

$$D(m,n) = \sum_{\alpha} \; [L_m(Q_m{}^{\alpha}) + L_n(Q_n{}^{\alpha}) - L_m(Q_n{}^{\alpha}) - L_n(Q_m{}^{\alpha})],$$

where $Q_m{}^{\alpha}$ and $Q_n{}^{\alpha}$ stand for the matrices showing dependence at direction α computed from the two textures respectively. For the first-order MRF textures, it suffices to use only the horizontal and vertical directions (i.e. $\alpha=90^{\circ}$ and $\alpha=180^{\circ}$). Therefore, by way of the SGLDM, representing the first-order parameters that serve to generate textures m and n by $\{b_{1m}, b_{2m}\}$ and $\{b_{1n}, b_{2n}\}$ respectively, we have:

$$D(m, n) = \Sigma \; [Q_m{}^{90}(i, j) - Q_n{}^{90}(i, j)]\{\ln [Q_m{}^{90}(i, j)] - \ln [Q_n{}^{90}(i, j)]\}$$

$$+ \Sigma \; [Q_m{}^{180}(i, j) - Q_n{}^{180}(i, j)]\{\ln [Q_m{}^{180}(i, j)] - \ln [Q_n{}^{180}(i, j)]\}$$

$$= 0.592(b_{1m}-b_{1n})[\ln((1+1.184b_{1m})/(1-1.184b_{1m}))$$

$$- \ln((1+1.184b_{1n})/(1-1.184b_{1n}))]$$

$$+ 0.592(b_{2m}-b_{2n})[\ln((1+1.184b_{2m})/(1-1.184b_{2m}))$$

$$- \ln((1+1.184b_{2n})/(1-1.184b_{2n}))].$$

To establish a Euclidean space, we need to further simplify the right-hand side of the above equation. Considering the Taylor expansion:

$$\ln[(1+x)/(1-x)] = 2x + 2/3\, x^3 + o(x^3),$$

when $|x| < 1$, we have

$$D(m,n) \approx C\, [(b_{1m}-b_{1n})^2 + (b_{2m}-b_{2n})^2],$$

where $C = 0.592*2*1.184$ is a constant. The maximum possible error introduced by this approximation is 17% when b_1 and b_2 are restricted to the range [-0.5, 0.5]. This error can be eliminated if we restrict b_1 and b_2 to a sufficiently small range.

We now consider the following definition of Euclidean n-space [15]:

If (1) X is a metric space,

    (2) f: $X \to R^n$ exists, where R^n is an n-dimensional real space, and

    (3) the metric d: X x X $\to$ R is generated by the norm $\| \bullet \|_2$,

then X is called Euclidean n-space.

Therefore, b_1 and b_2, in a neighbourhood of 0, approximately constitute two dimensions of a Euclidean space. If we use the NGLDM, which has been reported to give superior performance to the SGLDM in image classification tasks, we can use similar techniques to the above to derive the same approximate distance formula and therefore reach the same conclusion when b_1 and b_2 are chosen to be small enough so that their quadratic terms can be ignored.

We suggest that this space may be considered a perceptual texture space that can be used for visualisation purposes. This is because the definition of the distance metric of the Euclidean space is based on the results of texture classification that is consistent with human visual discrimination. The fact that the identical conclusions can be arrived at by using the SGLDM and NGLDM respectively also tends to show that our approach, albeit derived from two gray level dependence matrices, does not entirely depend for its perceptual validity on either of the two matrices' containing identical information to that extracted by the visual system.

However, our suggestion remains hypothetical and needs careful examination. The SGLDM and NGLDM classification schemes have been tested only on a small number of texture sets, the majority of which consisted of visually very distinct textures. We also recognise that even if the texture classification scheme underlying our established Euclidean space were more fully experimentally validated, our suggestion remains an hypothesis because the texture difference measure that we have chosen is not based upon strictly logical deduction from the classification scheme.

6. Demonstrations

The proposed texture space can be used for trial visualisations. We illustrate this usage by presenting several textures to show some functions of the planar coordinates (Figures 2-5).

Texture is likely to be able to enrich the information content in a display. Although texture variations are generally not perceived as accurately as colour or gray-scale variations, they are still capable of reflecting some data variations and reducing some ambiguity. Using textures may therefore be considered as a supplementary measure to visualising complex data when the sole use of other characteristics such as colour is not sufficient. Figure 6 shows the depictions of some realistic data by the MRF texture.

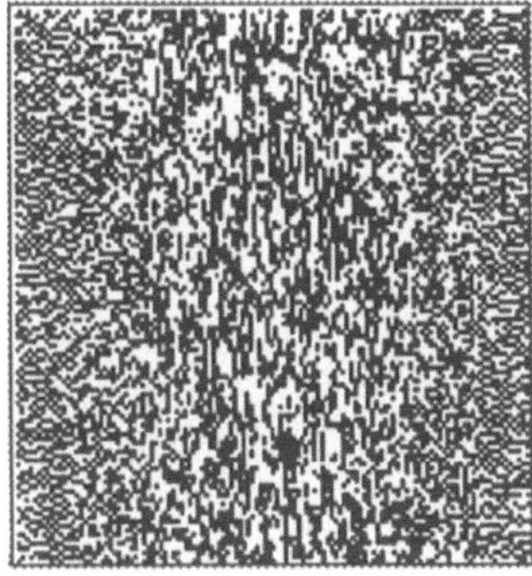

Fig. 2. Texture representation of
function f(x) = sinx, x ∈ [0, π].

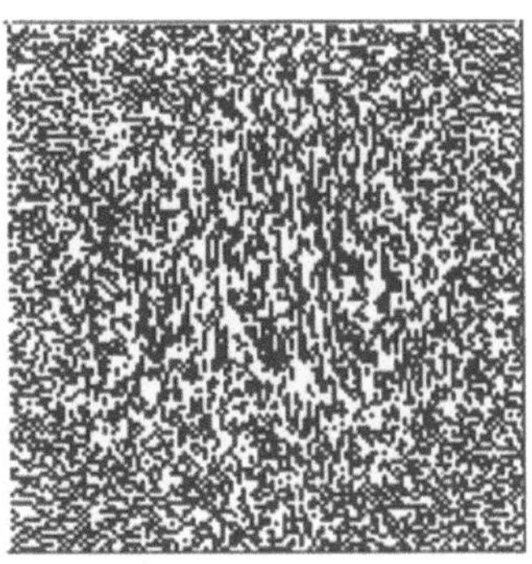

Fig. 3. Texture representation of function
f(x, y) = sinx + siny, x, y ∈ [0, π].

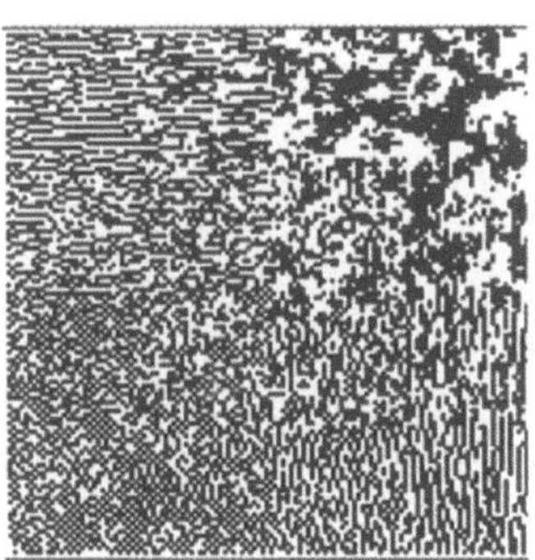

Fig. 4. Texture representation of
two-channel synthetic data:
Channel 1: f(x) = x;
Channel 2: f(y) = y.

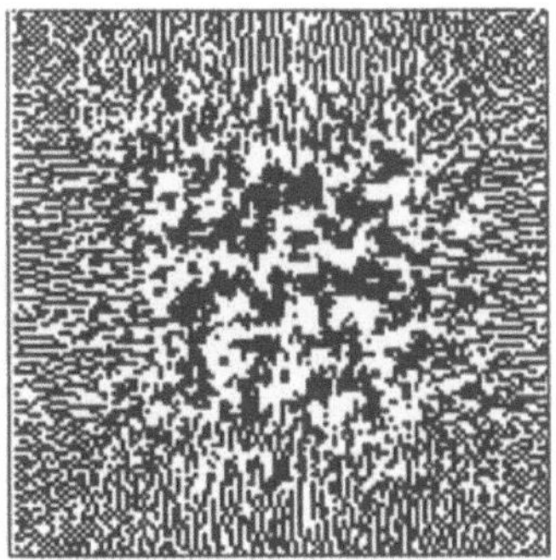

Fig. 5. Texture representation of
two-channel synthetic data:
Channel 1: f(x) = sin x, x ∈ [0, π];
Channel 2: f(y) = sin y, y ∈ [0, π].

The sampling algorithm used in our generation of MRF textures is the Metropolis algorithm, in which only the colour (gray-level) combinations that satisfy pre-defined proportions are possible. This algorithm therefore does not satisfy the positivity requirement in the definition of MRFs (which requires that any colour combination must be possible). Nonetheless, this algorithm works well empirically and, in addition, avoids the uni-colour tendency.

There are certain limitations or restrictions that must be considered beforehand, the major one being that only the low spatial-frequency information, or the smooth parts, can be effectively displayed.

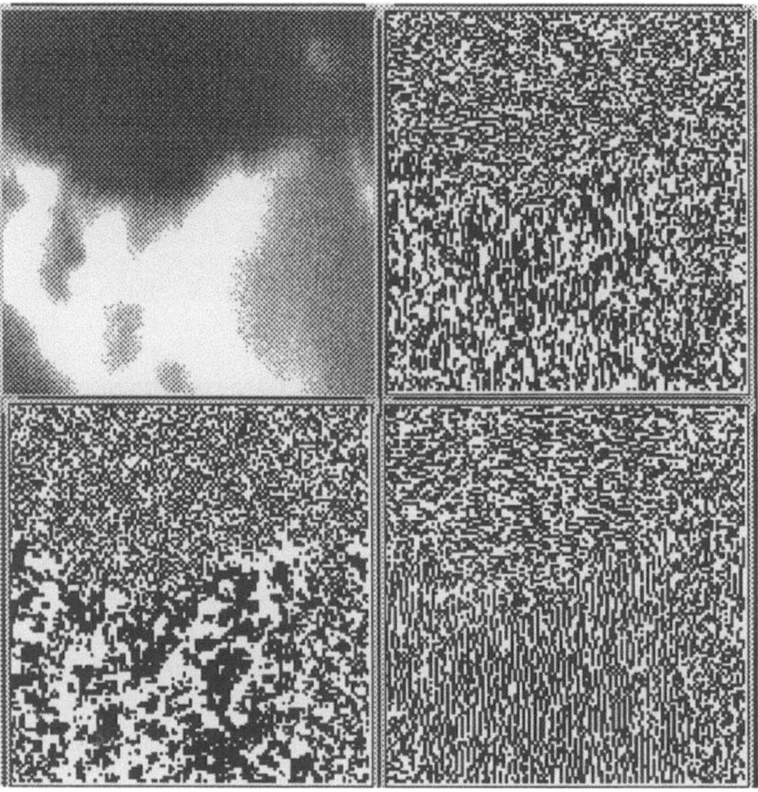

Fig. 6. Demonstration of using MRF textures to depict realistic data. The bottom-right gray-scale image is used as a reference. Different ranges of parameters (i.e. b1 and b2) are used to achieve different appearances in the other three images, which display the same data. In a colour display textures may be used to add (or enhance) some information channels to the existing image.

7. Summary

This paper has demonstrated that texture can be a visualisation medium. But there is still much to be understood about how to exploit texture representations in a systematic and controlled manner. We have chosen MRF stochastic textures to develop a systematic visualisation approach because the MRF has been intensively studied both in its theory and in its applications, and also because MRF sampling algorithms are inherently parallel. In this study, we have shown that two gray-level dependence matrices, the NGLDM and the SGLDM, of textures generated as samples of the first-order MRFs can be mathematically predicted by the corresponding MRF parameters. We have further suggested a texture difference measure based upon a classification scheme that uses those matrices, and through mathematical deduction established a Euclidean space for the first-order MRF textures, defining the distance metric by the suggested texture difference. Experimental evidence supports the perceptual success of these classification schemes. We therefore suggest that the proposed approach has advanced in the direction of gaining perceptual control over textures generated by varying MRF parameters.

On the other hand, we are aware of the lack of sufficient empirical results, notwithstanding our demonstrations, in support of our suggestion that the Euclidean space derived in this work may be used as a perceptual space for practical purposes. Our suggestion may be regarded as an hypothesis that requires experimental testing.

Acknowledgements

The authors wish to thank Dr. Paul Mackerras for his helpful comments on an earlier draft of this paper.

The first author has been supported by scholarships provided by the Australian National University and its Computer Science Department.

References

1. J.J. van Wijk: Spot noise texture synthesis for data visualization. Computer Graphics 25(4), 309-318 (1991)

2. C. Ware, W. Knight: Orderable dimensions of visual texture for data display: orientation, size and contrast. In: Human Factors in Computer Systems: CHI '92 Conference Proceedings. Monterey, California, May, 1992, pp. 203-209

3. G.R. Cross, A.K. Jain: Markov random field texture models. IEEE Transactions on Pattern Analysis and Machine Intelligence PAMI-5 (1), 25-39 (1983)

4. J. Yuan, T. Subba. Rao: Spectral estimation for random fields with applications to Markov modelling and texture classification. In R. Chellappa, A. Jain (eds): Markov Random Fields: Theory and Application. San Diego: Academic Press 1993, pp. 179-209

5. Yu. A. Rozanov: On Gaussian fields with given conditional distributions. Theory of Probability and Its Applications 12 (3), 381-391 (1967)

6. R.C. Dubes, A.K. Jain: Random field models in image analysis. Journal of Applied Statistics 16 (2), 131-164 (1989)

7. J.R. Berry, J. Goutsias: A comparative study of matrix measures for maximum likelihood texture classification. IEEE Transactions on System, Man and. Cybernetics 21 (1), 252-261 (1991)

8. R.M. Haralick, K. Shanmugam, I. Dinstein: Textural features for image classification. IEEE Transactions on System, Man and Cybernetics 3, 610-621 (1973)

9. L. Onsager: Crystal statistics I . A two-dimensional model with an order-disorder transition. Physical Review 65, 117-149 (1944)

10. G.F. Newell, E.W. Montroll: On the theory of the Ising model of ferromagnetism. Reviews of Modern Physics 25, 353-389 (1953)

11. M.S. Bartlett: Physical nearest-neighbour models and non-linear time-series II. Journal of Applied Probability 9, 76-86 (1972)

12. A. Gagalowicz: A new method for texture fields synthesis: some applications to the study of human vision. IEEE Transactions on Pattern Analysis and Machine Intelligence PAMI-3 (5), 520-533 (1981)

13. A. Treisman, S. Gormican: Feature analysis in early vision: evidence from search asymmetries. Psychological Review 95 (1) 15-48 (1988)

14. A. Treisman: Properties, parts and objects. In K. Boff, L. Kaufman, J. Thomas (eds.): Handbook of Perception and Human Performance Vol. 2: Cognitive Processes and Performance. New York: Wiley 1986, pp. 1-70

15. J.R. Giles: Introduction to the Analysis of Metric Spaces. Cambridge: Cambridge University Press 1987, pp. 1-7

A Multidimensional Multivariate Image Evaluation Tool

Pak Chung Wong   R. Daniel Bergeron
pcw@cs.unh.edu   *rdb@cs.unh.edu*

Department of Computer Science
University of New Hampshire
Durham, New Hampshire 03824

Abstract. Our current research focus is on the representation and visualization of multidimensional multivariate (mDmV) data. We are developing a visualization evaluation tool, whose primary goal is to provide an environment for visualization researchers to evaluate human responses to different computer generated visual images. It has the ability to create mDmV data with embedded stimuli, and display them in a variety of ways including icons. In addition, statistical analysis functions are provided for visualization researchers to study relationships among variates.

1 Introduction

Evaluating data visualization effects is a sophisticated process. Although a variety of tools exist for coding and supporting various visualization techniques, there are almost no tools that help visualization researchers to evaluate the *effectiveness* of these techniques. Off-the-shelf visualization packages such as AVS[1] could be used as the base for such an environment, but this would require substantial programming work.

In this paper, we describe a tool that aids visual image evaluation by automating the process. The evaluation process involves an examiner who produces a series of images, and a subject to respond to these images. The examiner can define the visualization technique to be used to display the data. This allows visualization researchers to evaluate the effectiveness of each technique.

The system is a Motif-based tool that makes extensive use of color and iconographic displays. It supports scientific visualization in a variety of ways:

- it generates mDmV test data with embedded stimuli whose characteristics are controlled by visualization researchers;
- it saves test data, in pre-defined format, as an ascii or binary file;
- it creates images from test data, and displays them either in color, or grey scale;
- it displays images with icons or plain pixels;
- it accepts keyboard and mouse responses from humans;
- it keeps track of the results and scores of each test session;
- it generates evaluation reports.

In the next section we describe the properties and the characteristics of our computer generated random data. In section 3, we describe the two major system components, their corresponding screens and basic system operations. Section 4 illustrates several data display mechanisms. This system supports statistics analysis among different variates, which we discuss in section 5. The issue of future enhancements is discussed briefly in section 6, and section 7 presents our concluding remarks.

2 Computer Generated Data

Data generated by the system consists of a white noise background with embedded stimuli in random locations. Each of these stimuli can have a different probability distribution. The system also supports multivariate stimuli with different probability distributions.

2.1 Random Number Generator

Anyone who considers arithmetical methods of producing random digits is, of course, in a state of sin. John von Neumann

The data created by the system needs a random number generator. The random numbers generated by a computer are called *pseudo-random* because they are obtained using deterministic rules actually stemming from a feedback system. The prevalent method is the linear congruential generator. The algorithm is based on a fundamental congruence relationship which can be expressed as

$$I_{i+1} = (aI_i + c)(mod\ m), \qquad i = 1, ..., n$$

where $a, c, m \in Z^+$. This recurrence is implemented in the random number generators such as *rand()* and *drand48()* in Unix libraries. The problem with this algorithm is that it is not free of correlation on successive calls. If k random numbers generated at a time are used to plot points in k dimensional space, the points do not tend to fill up the k-dimensional space.

Park and Miller[8] prove that the simple multiplicative congruential algorithm, defined by

$$I_{i+1} = aI_i(mod\ m), \qquad i = 1, ..., n$$

where $a, m \in Z^+$, can be as good as any of the more general linear congruential generators that have $c \neq 0$. The algorithm passes all theoretical tests for randomness with values of $a = 7^5$ and $m = 2^{31} - 1$. This algorithm includes a trick, based on an approximate factorization of m, to multiply two 32-bit numbers in any machine for which the maximum integer is $2^{31} - 1$ or larger. We chose this algorithm as the basis to generate our mDmV random data.

2.2 Random Variable and Probability Density Function

Before we continue the discussion of probability distributions, we need to clarify several statistics definitions.

For a given sample space $\mathfrak{S}$ of some experiment, a *random variable* is any rule which associates a number with each outcome in $\mathfrak{S}$. For example, with $\mathfrak{S} = \{Head, Tail\}$, a random variable X can be defined by

$$X(Head) = 1, X(Tail) = 0$$

The random variable X indicates a win (Head) or a loss (Tail) of a coin throwing experiment.

A random variable X is said to be *continuous* if its set of possible values is an entire interval of numbers. A *probability density function (p.d.f)* of a continuous random variable X is a function $f(x)$ such that for any two numbers a and b with $a \le b$,

$$P(a \le X \le b) = \int_b^a f(x)dx$$

In Figure 1, the probability that X takes on a value in the interval $[a, b]$ is the area under the curve of the density function.

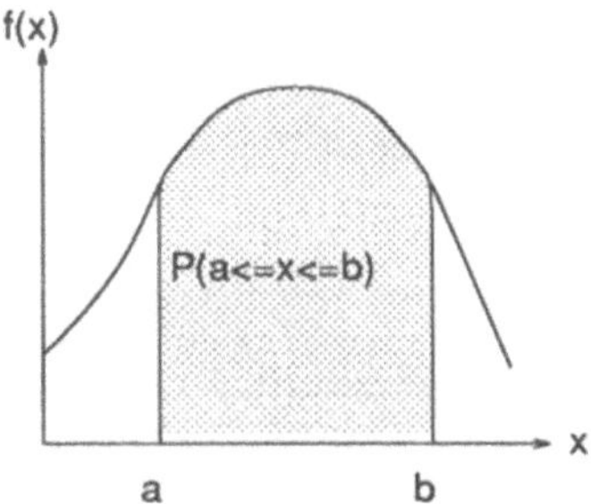

Fig. 1. $P(a \le x \le b) =$ the area under the graph of $f(x)$ between a and b.

2.3 Uniform Probability Distribution Background

The random noise background of each of the created data sets comes from a computer generated random data set with a uniform distribution. A continuous random variable X is said to have a *uniform distribution* on the interval $[a, b]$ if the p.d.f. of X is

$$f(x) = \begin{cases} \frac{1}{b-a} & \text{if } a \le x \le b \\ 0 & \text{otherwise} \end{cases}$$

as shown in Figure 2.

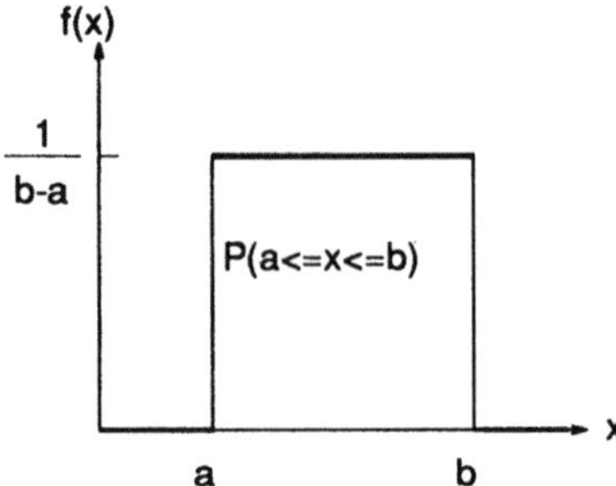

Fig. 2. A uniform distribution on the interval [a,b].

2.4   Non-uniform Probability Distribution Stimuli

In our test data set, non-uniform probability distributed stimuli are embedded in the white noise background in random positions. The most commonly used method to generate random noise data is the gaussian random generator. A continuous random variable X is said to have a *gaussian distribution* with parameters μ (expected value) and σ (standard deviation), where $-\infty < \mu < \infty$ and $\sigma > 0$, if the p.d.f. of X is

$$f(x) = \frac{1}{\sqrt{2\pi}\sigma} e^{-\frac{(x-\mu)^2}{2\sigma^2}} \qquad -\infty < x < \infty$$

The graph is symmetric about μ and bell-shaped. The center of the bell is both the mean and the median of the distribution. As depicted in Figure 3, large

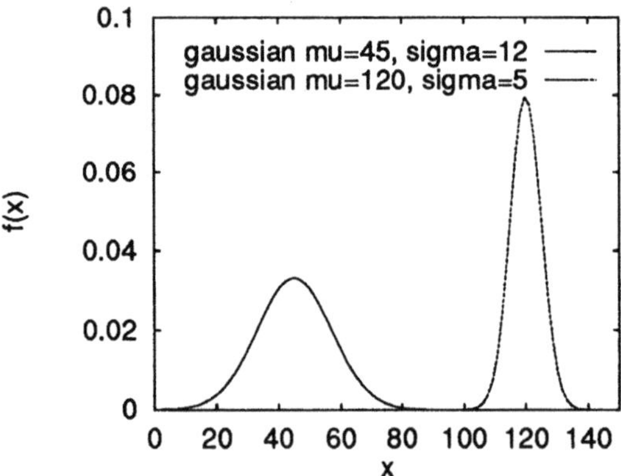

Fig. 3. Gaussian density functions.

values of σ yield graphs which are spread out about μ, whereas small values of σ yield graphs with a high peak about μ and most of the area under the graph quite close to μ.

There are efficient and accurate methods available for generating gaussian random numbers[11]. We use the Box-Muller method which applies the fundamental transformation law of probability[10]. (Refer to any standard text on numerical analysis for more details in this regard.) To create a gaussian distribution having μ_1 and σ_1 from a standard distribution, we need to calculate

$$Gaussian = \mu_1 + StandardGaussian() * \sigma_1$$

The gaussian distribution is taken as a model for a statistically healthy sample, which may not always serve our evaluation purpose. There are practical situations in which the variable of interest of the experiment might have a skewed distribution. Statisticians have developed another family of distributions, the *gamma distribution* as shown in Figure 4. For $\rho > 0$, the gamma function is

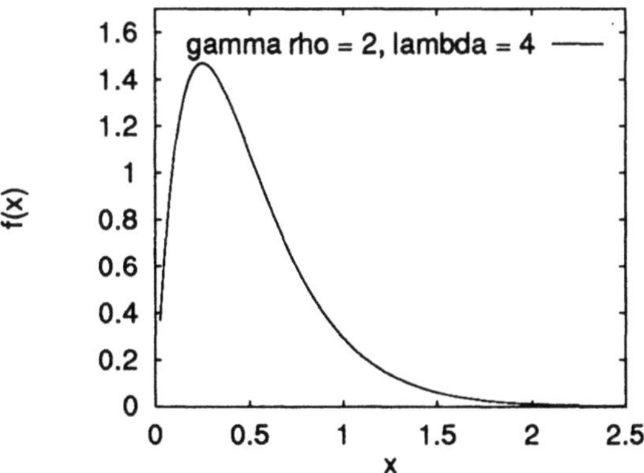

Fig. 4. A gamma density function.

defined by

$$\Gamma(\rho) = \int_0^\infty x^{\rho-1}e^{-x}dx$$

A continuous random variable X is said to have a gamma distribution if the p.d.f. of X is

$$f(x) = \begin{cases} \frac{\lambda^\rho}{\Gamma(\rho)}x^{\rho-1}e^{-\lambda x} & \text{if } x \geq 0 \\ 0 & \text{otherwise} \end{cases}$$

where $\rho > 0$ and $\lambda > 0$. The mean (μ) and standard deviation (σ) of a random variable X having the gamma distribution are

$$\mu = \frac{\rho}{\lambda}$$

$$\sigma = \frac{\sqrt{\rho}}{\lambda}$$

The parameter λ is used to control the spread of the gamma graph in the x direction. This is somewhat similar to the σ parameter in the gaussian graph.

Rubinstein[11] describes a total of eight different procedures to generate a random data set having a gamma distribution. We use an algorithm described by Knuth[7] that is valid for non-integer ρ, although it may not be necessarily the best algorithm for all values of ρ. Since our data are defined with σ and μ, a non-integer ρ is required in order to cover all values of σ and μ.

2.5 Gaussian versus Gamma

For larger values of μ (e.g., $\mu = \rho/\lambda = 4/1 = 4$), the gamma distribution has a typically bell-shaped form. Figure 5 shows gaussian and gamma functions that

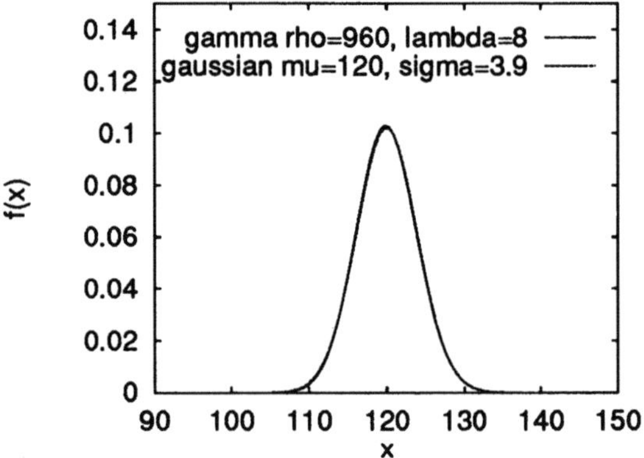

Fig. 5. Gaussian and gamma density functions with $\mu \gg 0$.

have the same values of μ and σ.

$$\mu_{gamma} = \frac{\rho}{\lambda} = \frac{960}{8} = 120$$

$$\sigma_{gamma} = \frac{\sqrt{\rho}}{\lambda} = \frac{\sqrt{960}}{8} \simeq 3.9$$

The symmetric bell-shaped gaussian curve almost matches the gamma curve.

However, the shape of the gamma graph is altered drastically by a change of the value of μ. It starts showing the skewness as the value of μ approaches the origin. In Figure 6, both graphs have the same values of σ and μ. The gaussian graph keeps its symmetric bell-shaped curve. It actually intersects the y-axis and continues on the negative side of the axis. The gamma graph, however, has the density going to zero as it approachs the origin. This property is important to our study because we are also interested in evaluating data that is constrained to certain ranges, but is not distributed symmetrically in those ranges. The gamma distribution is not perfect. The peak value of the curve is not always equal to μ, and the zero density property only occurs at the origin end. To generate data with zero density on both ends, we have to bisect the data range into two halves and map the data from the left side (closest to the origin) to the right side of the peak value.

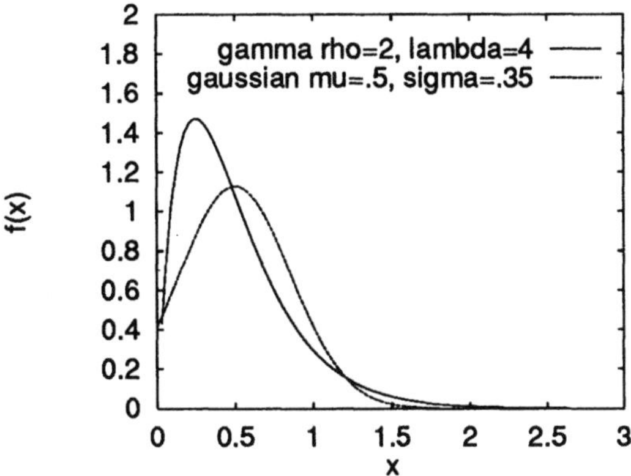

Fig. 6. Gaussian and gamma density functions near the origin.

When the gamma distribution graph has $\rho = 1$,

$$f(x) = \begin{cases} \lambda e^{-\lambda x} & \text{if } x \geq 0 \\ 0 & \text{otherwise} \end{cases}$$

which is just the exponential distribution, as shown in Figure 7. The exponential

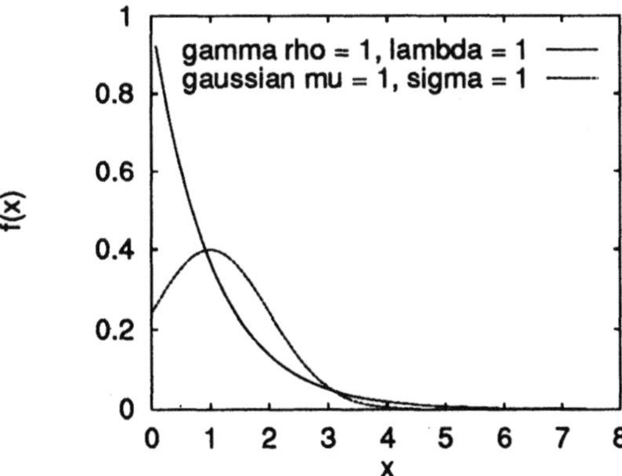

Fig. 7. Gaussian and gamma density functions as $\mu \to 0$.

distribution graph $f(x)$ is strictly decreasing as x increases from 0 with

$$\mu = \frac{1}{\lambda}$$

$$\sigma = \frac{1}{\lambda}$$

Although the exponential distribution is a special case of the gamma distribution, it lacks the characteristics of a typical noise function which rises from 0 to a maximum and then decreases. We do not support gamma distributions with $\rho = 1$.

The statistical properties of these probability distributions are not the most important issue in our design. We are looking for random data sets with certain locality patterns and data correlations. The data set must have positive and negative errors around a certain target. The use of the gaussian, the gamma, or any other distributions such as the chi-squared, provides us with an approximation of our desired data.

3 System Components

The system consists of two separate modes: *data generation* and *experiment monitoring*. The data generator is used by the examiners who define experiments, whereas the experiment monitor is used by subjects.

3.1 Data Generation Component

The data generation screen is designed to help visualization researchers define their data. Figure 8 shows that the system contains a number of field widgets. In

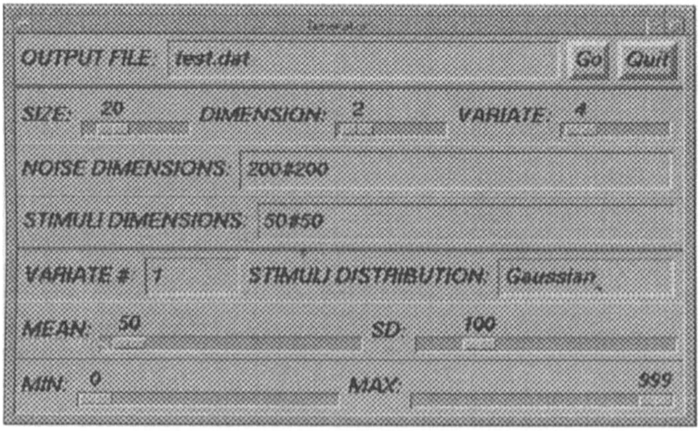

Fig. 8. Data generation screen.

the top pane, the *OUTPUT FILE* widget allows a user to specify the output file name. The *GO* widget creates a new data file defined by the other widgets. The *QUIT* widget exits the system. In the middle pane, the *SIZE* widget contains the number of image frames to be generated. The *DIMENSION* widget specifies the dimensions of each image and its stimulus. The user can define the number of variate with the *VARIATE* widget. The *NOISE DIMENSIONS* and the *STIMULI DIMENSIONS* widgets accept character strings containing dimension values. For example, to specify a 50x50 2-D image, the user enters "50♯50" in the field. The ♯-sign serves as delimiter for each dimension. The characteristics of each variate is defined at the bottom pane. The user specifies the variate number in the *VARIATE* ♯ widget. The distribution type, the mean, the standard deviation, the minimum value, and the maximum value of the data set can be described with the *STIMULI DISTRIBUTION* widget, the *MEAN* widget, the *SD* widget, the *MIN* widget, and the *MAX* widget. The default values of the *MAX* and the *MIN* widget fields are 255 and 0, which are the color range or gray scale range of most graphics terminals. The value of the *MEAN* widget has to be between the values of *MAX* and *MIN*. The maximum values of *SD* is one half of the *MAX-MIN* range. A similar shell command line version of this data generation program is also available.

3.2   Experiment Monitor Component

The second component is a pair of experimental target screens: display definition screen and data display screen.

The display definition screen is shown in Figure 9. The *DATA FILE* widget

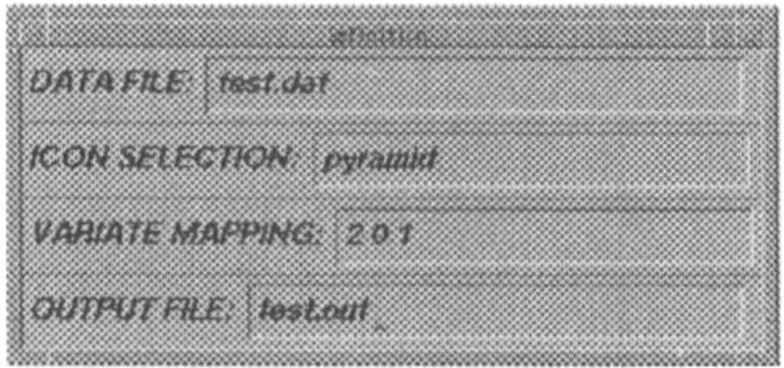

Fig. 9. Display definition screen.

and *OUTPUT FILE* widgets allow the examiner to define the input and the output filenames. The examiner has a choice of different icon display techniques provided by the system. Some of these techniques have a fixed number of parameters because they are designed to support certain domains of mDmV data. There are also general purpose icons that accept a variable number of variate-parameter mappings. This mapping can be specified by a character string entered in the *VARIATE MAPPING* widget. Consider a trivariate data set; a mapping string of *"2 0 1"* maps the third data variate to the first icon parameter, first variate to the second parameter, and the second variate to the third parameter. The system itself does not contain any icon generators. It is supported through the Flexvis[3] system being developed at the University of New Hampshire.

The data display screen is intended for novice keyboard users. There are two major panes in this system, as depicted in Figure 10. The *control pane* at

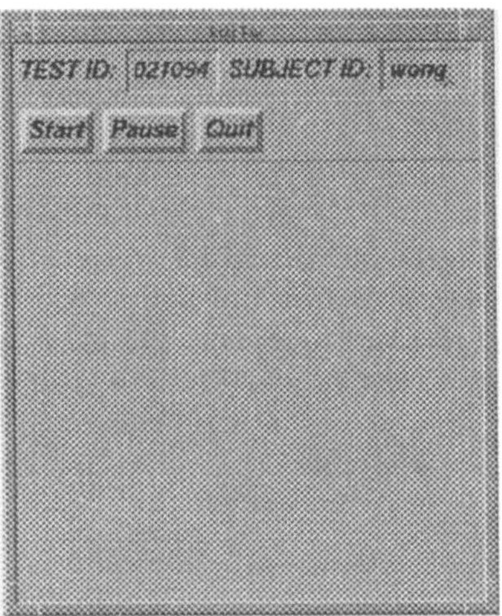

Fig. 10. Data display screen.

the top allows subjects to enter their identifiers, and to communicate with the system. The *workspace pane* at the bottom displays the target images created in the previous component. The subject uses the mouse pointer to answer the questions in the control pane. He/she only needs to know three operations of this system: press the start button to start; press the pause button to stop; and point-and-click the approximate locations of the stimuli in the workspace pane.

4   Data Displays

The system provides a default data display using individual pixels to represent the value of the mDmV data. The values of each variate is mapped to a true colormap with 256 colors. The variates of each dimension point are then arranged together and displayed as a unit in the image as shown in Figure 11. This method supports the most fundamental mDmV display. But it is our interest to use this tool to evaluate images with various user-defined iconographic displays. Figure 12 shows the use of pyramid icons[3] to display a different mDmV data set with a relatively larger embedded stimulus in the lower right quadrant.

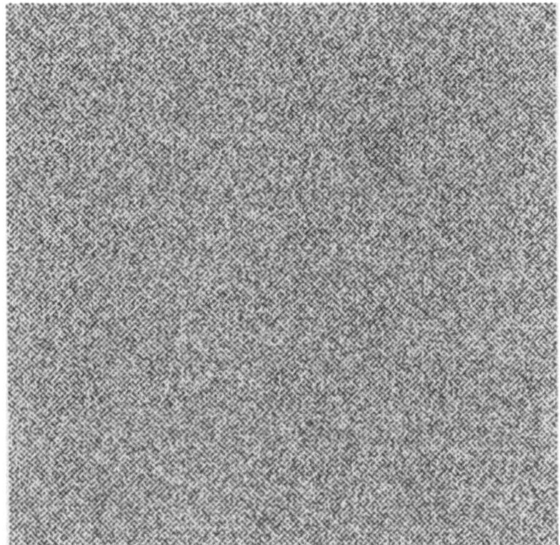

Fig. 11. A simple pixel display of an mDmV data set.

Fig. 12. An iconographic display of an mDmV data set.

4.1 Iconographic Displays

A number of researchers[2][5][9] have experimented with iconographic displays
in which each multidimensional point is represented by an icon which changes
shape, color, etc. based on the values of component variates. When the sample
points are relatively dense, the iconographic display presents texture patterns
that vary according to the nature of the data. The proper selection of icon and
variate mapping can produce visual patterns that represent important multidi-

mensional relationships among the variates. The individual icons themselves are not intended to be discernible, but the overall texture pattern that results can show areas in which the combinations of the parameters appear to be interesting and thereby can be used to focus the application of other analysis techniques.

All of the mentioned researchers have shown promising results with specific image examples. However, none have proven that a specific display technique is particularly effective for specific domains. It is our goal to provide a mechanism to study the effectiveness and usefulness of various visualization techniques.

4.2 Presentation Issues

Some preliminary observations have been recorded since the first version of the system. One of the interesting results is that motion changes are more perceptible than we expected. When one image frame is being replaced with a new one, the flicking motion actually brings out the location of the embedded stimuli.

Although the motion change factor is interesting, we would like to provide an environment in which all the possible underlying factors are well under control. That means we have to eliminate the flicking. Fortunately, this problem can be solved simply by inserting a blank screen in between images of the display sequence.

A second solution attacks the source of the problem, the pre-defined fixed white noise background. Instead of using one single background pixmap, multiple backgrounds are generated and applied randomly for display. Now the screen swapping involves more pixel exchanges. As a result, the extraneous perception cue is substantially eliminated.

5 Statistical Analysis

One of the important goals of our research is examine the relationships among variates. We want to learn how humans respond to texture patterns found in data displays. To do this, we need to control the statistical relationships among variates.

5.1 Covariance and Correlation

Consider any two variates (x and y) of mDmV data. We want to know the relationship between these two variates at all points, i.e., (x_1, y_1), (x_2, y_2), ..., (x_n, y_n). If large x's are paired with large y's, we say x and y has a positive relationship. Similarly, if large x's are paired with small y's, x and y has a negative relationship. This relationship can be described as:

$$\sum_{i=1}^{n}(x_i - \bar{x})(y_i - \bar{y})$$

where $\bar{x}$ and $\bar{y}$ are the mean of x and y. This measurement of covariance, however, has a serious defect: its computed value depends critically on the units of measurement. Instead, we use the sample correlation coefficient to represent the relationship between two variates. The sample correlation coefficient r for n pairs of $(x_1, y_1), ..., (x_n, y_n)$ is defined as:

$$r = \frac{\sum (x_i - \bar{x})(y_i - \bar{y})}{\sqrt{\sum (x_i - \bar{x})^2} \sqrt{\sum (y_i - \bar{y})^2}}$$

To eliminate $\bar{x}$ and $\bar{y}$ for efficient computation, the equation is modified into:

$$\frac{n \sum x_i y_i - \sum x_i \sum y_i}{\sqrt{n \sum x_i^2 - \left(\sum x_i\right)^2} \sqrt{n \sum y_i^2 - \left(\sum y_i\right)^2}}$$

The correlation coefficient, r, between two variates (x and y) is said to be the *observed* correlation coefficient. It represents the strength of the relationship of these two variates in an mDmV data set, and is usually used to make inferences about the *population* correlation coefficient, R, between the population X and Y. In our data generation component, R is used as a correlation constraint to produce our test data.

5.2   Joint Probability Distributions

A joint probability distribution $f(x, y)$ of a normal population is called the *bivariate normal probability distribution*. It can be specified by the following equation:

$$f(x, y) = \frac{1}{2\pi \sigma_1 \sigma_2 \sqrt{1 - R^2}} e^{-[(\frac{x-\mu_1}{\sigma_1})^2 - 2R(\frac{x-\mu_1}{\sigma_1})(\frac{y-\mu_2}{\sigma_2}) + (\frac{y-\mu_2}{\sigma_2})^2]/2(1-R^2)}$$

where $-\infty < x < \infty$, $-\infty < y < \infty$, μ_1 and σ_1 are mean and standard deviation of X, μ_2 and σ_2 are mean and standard deviation of Y, and R is the population correlation coefficient of X and Y. To generate a random data set having a bivariate normal distribution is rather complicated, it can be done by assigning the appropriate correlation matrix. We are currently working on a random number generator that accepts correlation constraints between variates. Like the current version, both gaussian and gamma distributions are supported.

6   Future Enhancements

In mDmV visualization research, it is believed that the data can be better understood when all dimensions and all variates are presented visually as one display. Our current system is basically a 2DmV system. Higher dimensional displays are necessary to satisfy visualization research needs. We also plan to add the capability to generate mDmV data with cross-correlations[6] of more than two variates.

A database can play an important role in our system. Not only can it maintain the results of the evaluation process, but also the data and the images themselves. We would like to include a database management system in our system eventually.

7   Concluding Remarks

This image evaluation tool enables visualization researchers to study human responses to mDmV images. This is the first step towards a complete representation of higher order images, in terms of the syntax of the image representation as well as the semantics among them.

Acknowledgements

The authors' work has been supported in part by the National Science Foundation under grant IRI-9117153.

References

1. *DEC AVS: User's Guide for Ultrix Systems*, DEC, May 1992.
2. Beddow, Jeff, "Shape Coding of Multidimensional Data on a Microcomputer Display," *Proceedings IEEE Visualization '90*, San Francisco, Oct. 1990, IEEE Press, pp. 238-246.
3. Calder, Bartley H., "An Interactive Scientific Visualization Application Development Environment," Computer Science Technical Report 91-09, University of New Hampshire, May 1991.
4. Devore, Jay L., *Probability and Statistics for Engineering and the Sciences*, Brooks/Cole, Monterey, California, 1987.
5. Grinstein, Georges G., Ronald M. Pickett, and Marion Williams, "EXVIS: An Exploratory Visualization Environment," *Proc. Graphics Interface '89*, CIPS, Toronto, 1989, pp. 254-261.
6. Jungster, Jerome, "A Statistical Impetus For Development of Scientific Visualization Techniques," Technical Report, Institute for Visualization and Perception Research and Department of Computer Science, University of Massachusetts at Lowell, Lowell, MA, April 1992.
7. Knuth, Donald R., *Seminumerical Algorithms, The Art of Computer Programming*, 2nd edition, vol. 2, Addison Wesley, Reading, MA, 1981.
8. Park, Stephen K., and Keith W. Miller, "Random Number Generators: Good ones are hard to find," *Communications of ACM*, vol. 31, pp. 1192-1201, Oct 1988.
9. Picket, Ronald M., and Georges G. Grinstein, "Iconographics Displays for Visualizing Multidimensional Data," *Proc. IEEE Conf. on Systems, Man, and Cybernetics*, Beijing, May 1988.
10. Press, William H., Saul A. Teukolsky, William T. Vetterling, and Brian P. Flannery, *Numerical Recipes in C, The Art of Scientific Computing*, 2nd edition, Cambridge University Press, 1992.
11. Rubinstein, Reuven Y., *Simulation and the Monte Carlo method*, Wiley Series in Probability and Mathematical Statistics, John Wiley & Sons, 1981.

Visualizing Electromagnetic Data

Janet Haswell*

Rutherford Appleton Lab., Oxfordshire. ENGLAND

Abstract. This paper looks at how a commercial visualization system was used to produce interactive visualizations of simulated Eletromagnetic data. By firstly looking at techniques available for viewing vectors on an unstructured grid, an idea is given of the weaknesses in the system. The user of the simulation package takes advantage of symmetry within the problem to reduce the amount of computation required. Rather than duplicate unstructured grid data, a simple mirror tool has been developed to duplicate geometries instead.

Finally, a description is made of how the techniques were used to perceive the changing flow in time-varying 3D vector field data.

1   Introduction

As part of an evaluation in late 1991 to early 1992, a sample 3D time-varying dataset was obtained, courtesy of Vector Fields Ltd, to test the capabilities of visualization systems and to determine whether such a system would be of use to engineering application developers. Four systems were considered suitable as application builders: AVS, apE, Khoros and Explorer (the latter was not officially released at the time). The initial findings were reported as a case study in [5]. At the time, AVS proved to be the only system system capable of processing unstructured grid data directly. Explorer 1.0 did have the more general-purpose hierarchical data format'Pyramid', but few tools within Explorer could process Pyramid formats. In particular, no tools were available to handle vector-based Pyramid data.

The initial case study has been further developed to take account of improvements in AVS and more complete information on the nature of the data.

To develop an application in AVS (see [1], [2] and [3]) the user connects together building blocks, called Modules, into a Network. Two public domain AVS modules (obtained from the International AVS centre) were used to produce some of the images in this paper:

cone developed by Scott Lamson from General Electric, converts lines into cones to show vector magnitude and direction. This can threshold vectors below a selected magnitude.

harwell colormap developed by Kevin White (while at Stardent Computer Ltd.) and Ian Currington from AVS Inc. (now AVS/UNIRAS Ltd.). This generates gamma corrected colourmaps.

* Many thanks go to Dr. C.R.I. Emson from Vector Fields Ltd., for supplying the data and for feedback on the visualizations

A Silicon Graphics Crimson, VGXT with 64MBs of memory, running the visualization package, AVS release 5.0, was used in the production of images and animations.

2 The Data

The visualizations produced for this report came from a single data set, produced by a commercial electromagnetic field analysis package, ELEKTRA [4], supplied by Vector Fields Ltd., Oxford, UK. This package analyses electric and magnetic fields of general magnetic devices to produce solutions at a finite number of points (nodes) through 3D space.

Vector Fields also have a post processing package, OPERA [4], which takes the solution and produces static views. Rendering and viewpoint transformations are all performed in software to convert the 3D objects into a 2D image. Each minor change, such as a translation of objects, results in the recalculation of hidden surfaces. This makes animations difficult and time-consuming to produce.

The problem under investigation involves a magnetic coil with shielding material, surrounded by air. An electric current, following a sinusoidal cycle, flows just outside the magnet to induce a magnetic flow.

The simulation modelled a section of a toroidal magnet with the air surrounding it. The data consisted of: an unstructured set of cells (finite elements), material identifiers for each cell, 3D complex vectors at each node/point (at the "corners" of the elements) in the mesh. For this data, the cells were all hexahedral. Although it is not unusual for a model to contain up to 200,000 cells, the sample problem contained only 864 cells with 1090 nodes.

It was then possible to compute intermediate vectors in the flow cycle, using the real (R) and imaginary (I) components using the equation:

$$\cos\theta\bar{R} + \sin\theta\bar{I}, \qquad 0^\circ \leq \theta \leq 360^\circ$$

where $180^\circ \leq \theta \leq 360^\circ$ mirrors the vectors in the range 0° to 180° (i.e. the direction of the vectors are reversed).

An AVS reader was developed to import the data from an ASCII representation into the AVS unstructured cell data (UCD) format. Since vector magnitude appeared to be important for this data, the magnitude of the real and imaginary vectors were computed for each node. In addition to the real and imaginary components, space was reserved for a single intermediate solution - the user selects the number of degrees from the cycle, then the reader recomputes the 3rd set of data.

Although the model includes a description of the current carrying coils, these are independent of the mesh and so were not processed by the AVS reader. Within OPERA, the coils are converted to polygonal representations and so it would be possible to extract this section of code and generate AVS geometry (polygonal) data.

3 Visualization techniques for flow data

A good overview of possible techniques for flow visualization can be found in [6]. The number of modules in the AVS networks used for this section varied in size from 7, for the simpler illustrations, to over 30. To explore the flow techniques, only the real vector component was used to produce the illustrations.

3.1 Viewing the Mesh

The first step in understanding the data is to get an informative view of the mesh. Three typical orthographic views are shown in Figure 1 but only view c) helps to see the regular pattern used in generating the grid. This regularity would not normally occur with completely unstructured grids.

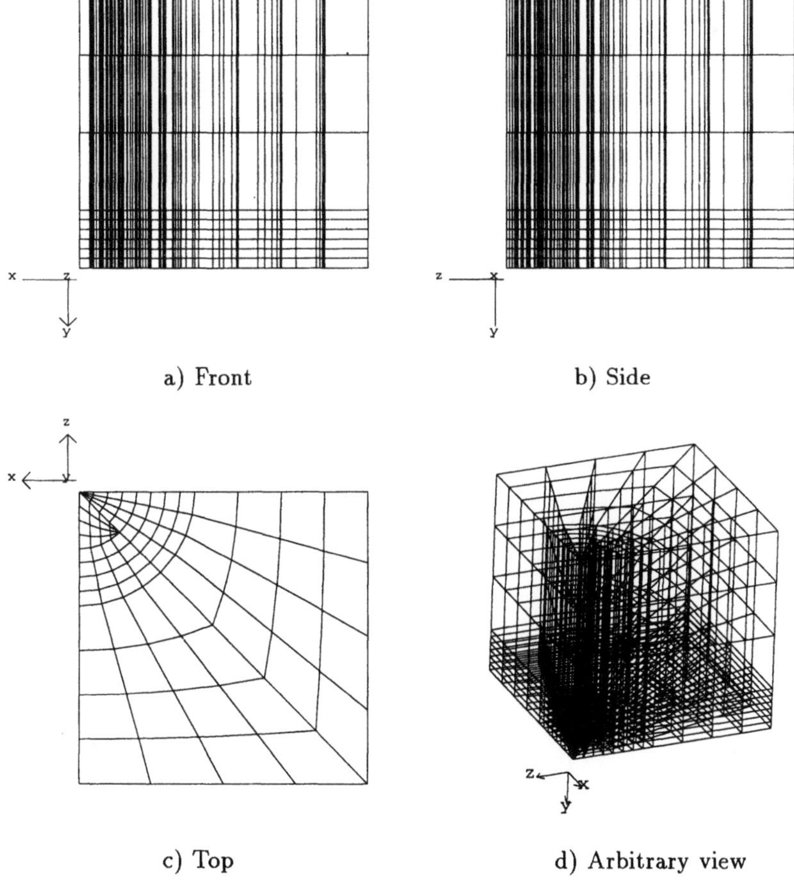

Fig. 1. Orthographic projections of the mesh.

Particularly when animating, orthographic views can confuse the user as it becomes difficult to distinguish foreground and background objects. In preference to orthographic projections, the default perspective projection in AVS gives reasonable results, see Figure 2. It is possible to vary the amount of detail shown for the mesh, from showing every line throughout the grid, to an outline of the structure only. In all cases displaying all lines for this data proved too cluttered but does give an idea of the complexity of the underlying grid. In some cases even the surface mesh interfered with the visualization, in particular with animations and arrow plots, and so the surface mesh was reduced to an outline of the 3 materials present, Figure 2c.

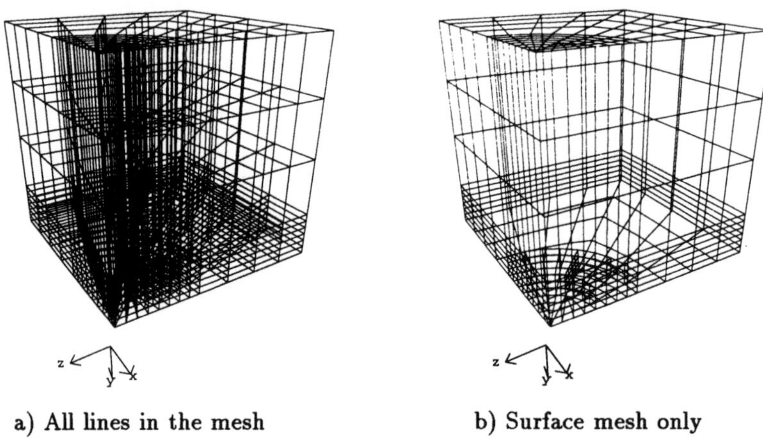

a) All lines in the mesh   b) Surface mesh only

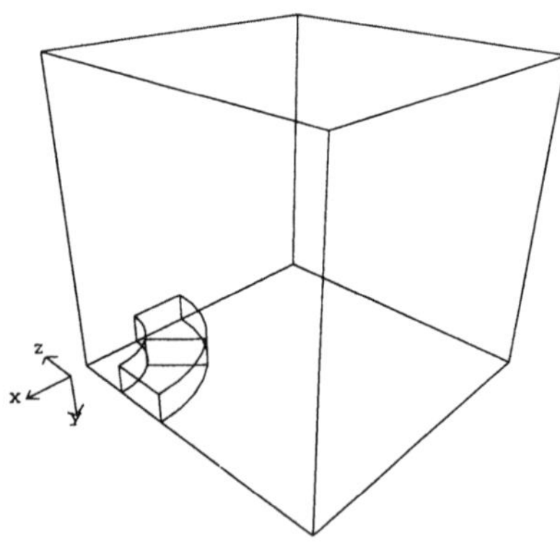

c) Outline of the materials

Fig. 2. Perspective projections of the mesh

After a short time interacting with perspective renderings, the orthographic version appeared more difficult to perceive both for wireframe and shaded views. Faces appear to be larger further away than those nearer when no perspective was present.

It is possible in AVS to shrink the finite element cells, Figure 3. This too can give a useful overall view of the grid structure. Using crop (by location) and threshold (by data value) it can also reveal internal patterns.

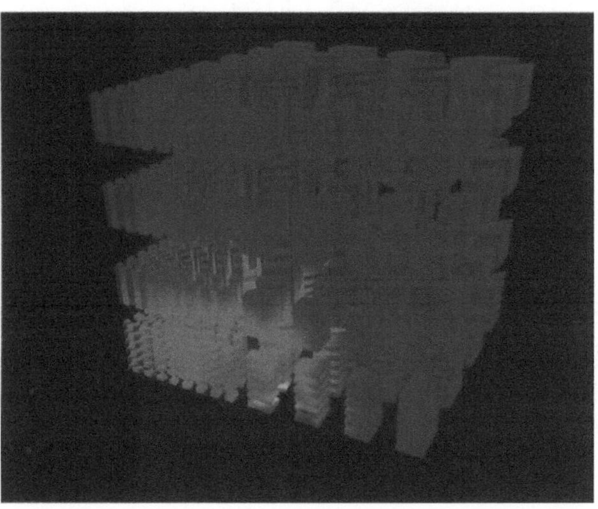

Fig. 3. Finite element cells shrunk by 50% to partly view internal structure of the grid

3.2 Using Arrows to Represent Vectors

An easy representation in AVS is to display the vectors as arrows, Figure 4. Since a large number of vectors were to be shown the outline of the materials was used to reinforce their position. The magnitude of the vectors was ignored as larger vectors occur in the already cluttered region of the magnet.

It is possible to crop and threshold by cells but the vector data is node-based so cannot be reduced in a similar manner, although since the cell format relies on nodes this should not be the case. Instead it is possible to select a region of interest and interpolate vectors onto a regular grid - but this would result in losing the fine grained resolution of an unstructured grid.

Instead, an indirect route was used to crop the geometrical object representing the vectors rather than the data itself. Figure 5 shows an example where the vectors are cropped by 3 planes, and the camera is zoomed in to see more detail. Similarly, additional planes could be used to crop away more regions. Although not immediately obvious from a static picture, fewer vectors are shown while other objects in the scene are complete (such as the outline mesh). In addition depth cueing could be used on the screen to distinguish between foreground and

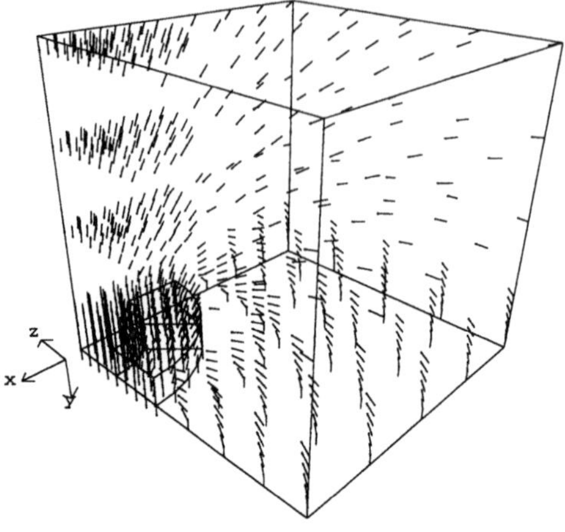

Fig. 4. Vectors displayed as arrows

background objects. The length of the arrows indicate relative magnitude. Since
the view concentrates on a small region, the large vector lengths can be scaled
to a size sufficiently small to clearly see all the detail.

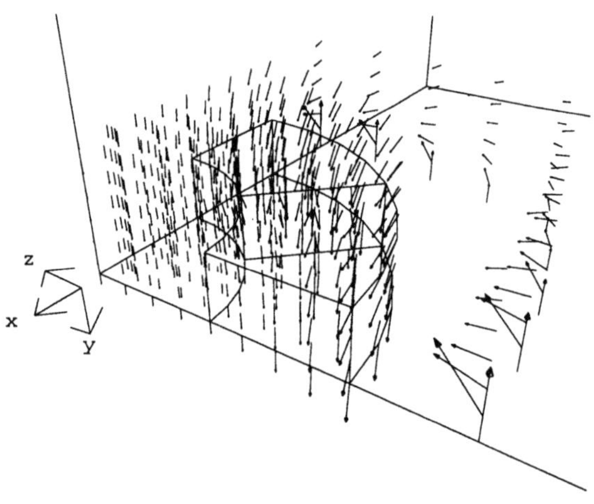

Fig. 5. Vectors cropped by location

3.3 Cones to Represent Vectors

Arrows, even with 3D heads, do not give a reasonable sense of direction. Using
a public domain module to convert the vector lines to cones gave better results.

Since it was possible to threshold small vectors it was possible to concentrate on larger magnitudes. The Vector Fields post-processor produces cones with red for external faces and yellow for internal faces to make it more obvious whether a vector is facing towards or away from the viewer. This technique was simulated using two sets of cones, a yellow set scaled slightly smaller than the red, Figure 6.

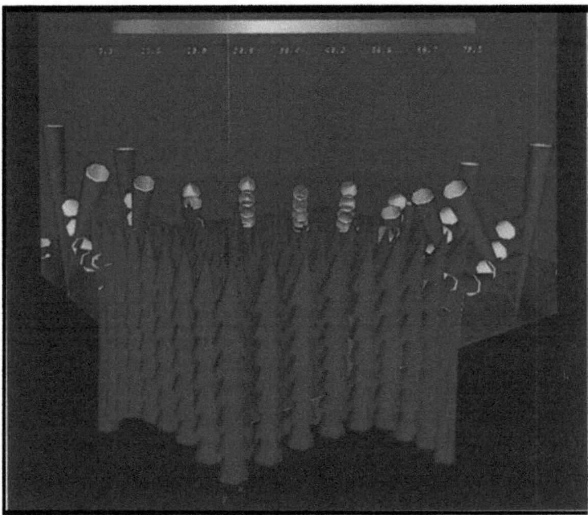

Fig. 6. Two-toned cones reinforce vector direction

3.4   Field Lines

Using arrows or cones did not give a good overall picture of the magnetic field. Instead streamlines were tried, Figure 7.

No variable line widths were possible on the screen with AVS, although it is possible to set line widths when converting wire-frame objects to PostScript line-drawings. Single pixel wide lines on a system without anti-aliasing do not produce satisfactory results - on photographic output, sections of a line could appear fainter, and in animations recorded onto video, lines flicker between varying intensities. To overcome this limitation, lines were converted into tubes, as in Figure 8.

Flux tubes would be a better representation, but without access to the source code of the tool which converts UCD vectors to streamlines this would not be easy and would involve re-inventing existing software. Access to the tube filter would be more likely and could be extended to use the reciprocal of the vector magnitude to give tube thickness along the field. However this would not take account of any twist(s) in the flow.

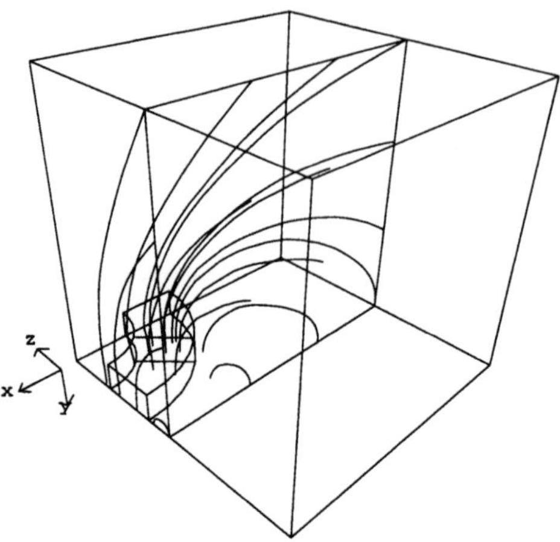

Fig. 7. Magnetic field lines for the real component vectors

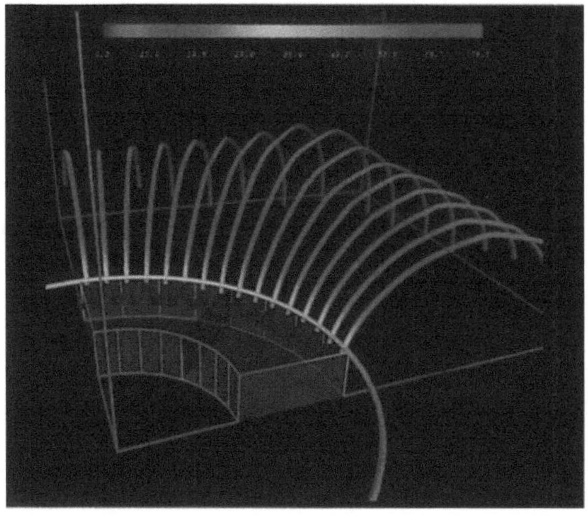

Fig. 8. Streamlines, replaced by tubes

Using stream ribbons can make twists in flow obvious, but for this data the main advantage was to emphasise the magnitude of the vector along the flow. In Figure 9 three circular probes are positioned to lie (a) near the coil carrying the electric current, (b) close to the outer edge, and (c) close to the inner edge of the magnet. For clarity, the field lines close to the magnet are represented as ribbons while the flow near the coil as lines.

In order to reinforce the perception of the strength of flow, a vector representation (such as field lines) could be combined with a volume rendering of the magnitude of the vectors throughout the volume. Although volume rendering,

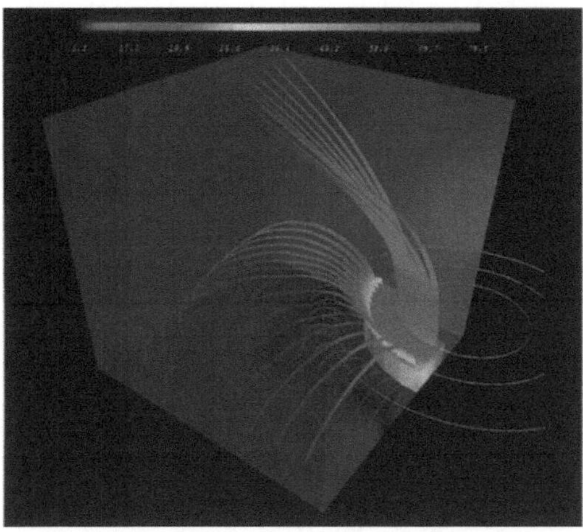

Fig. 9. Three circular probes give distinguishable sets of streams

i.e. of scalar components, is possible in AVS for UCD data, it is not possible to combine this rendered image with polygonal data. From release 4.0, AVS allowed mixed volume and geometry rendering for uniform 3D grids. A future release may do the same for UCD too. It is possible to set the degree of transparency for a geometry object from within the AVS geometry viewer. A fair approximation to volume rendering was obtained by rendering the section of magnet opaquely with the faces of each air cell transparent. Even with a modest number of cells the view looked very complicated and the field lines were difficult to see. Instead just the outer surface of all the air cells resulted in a much clearer visualization.

The polygonal representation of the stream ribbon appears to be a triangular strip. It is possible that the centre and edges of the strip could be determined and then a circular sweep, by less than $180°$ would result in a tube with 2 split sides. The splits would show twists in the flow. It would be considerably easier if standard AVS modules could be tailored and extended.

3.5 Using a Mirror Tool to View the Whole Problem

To simplify and reduce the calculations in a complex flow it is often necessary to rely on symmetry. It may be possible to look at half, quarter or even an eighth of the overall problem. When attempting to visualize the results, by just viewing this fraction of data it may be difficult to see what is happening - the overall flow pattern may not be obvious. Duplication of the solution data, with appropriate transformations, could be used. Since the original data could be large to begin with, this would add unnecessary overheads. Instead an AVS module was developed to duplicate the geometric representations of the data, which proved considerably more efficient. This module allows the user to select mirrors in x and/or y and/or z, with/without combinations.

For the given dataset the model was halved along the plane x=0, then y=0 then z=0 so the simulation relied on an 8-way symmetry. By selecting this new module to mirror in x, y, z, plus combinations, the streamlines shown in Figure 7 are transformed into a more obvious flow pattern, Figure 10. Using the mouse to interactively rotate and tumble this representation greatly increased the user's 3D perception. If a small number of seed points were used, it was also possible to re-position the streamline probe and get near real-time updates showing the mirrored field lines from the new locations.

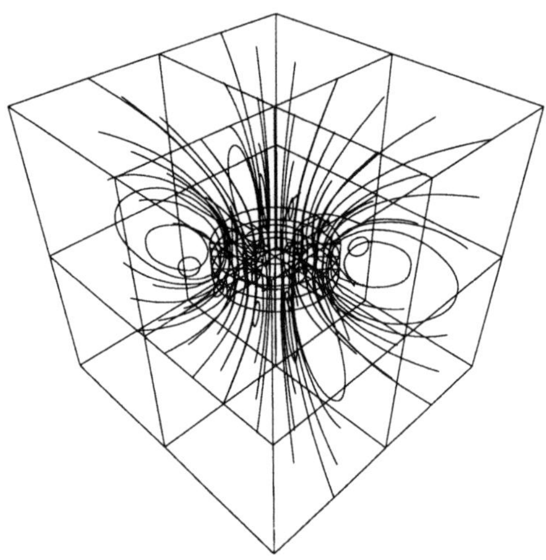

Fig. 10. Streamlines, duplicated and mirrored to give the overall picture

Two windows were used with the mirror tool. The first showed the orginal section of data and was used to interactively reposition probes, such as streamlines. The second showed the mirrored objects, an option in the mirror tool allows objects with 'probe' in their name to be hidden so the streamline probe would not appear in this view.

The mirror tool allows the user to select transparency levels on the objects being mirrored. Figure 11 shows stream ribbons mirrored, with the transparent surface of the air coloured by vector magnitude. Since several transparent faces overlap, this could not be rendered accurately in graphics hardware on an SG Crimson VGXT and instead the AVS software renderer was used.

By using the mirror tool to mirror the air geometry in x, y, and z only, it was possible to get a clearer idea which field lines were directed towards and away from the user, Figure 12. This technique had the added advantage of producing a reasonable rendering in hardware on the Crimson and so could be rotated in real-time.

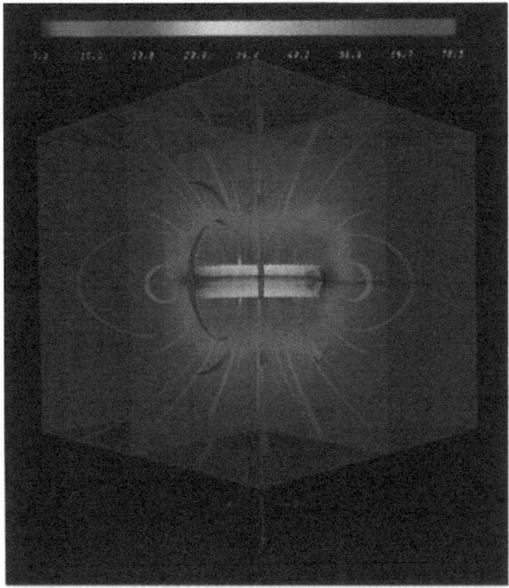

Fig. 11. The complete magnet, stream ribbons, and surrounding air

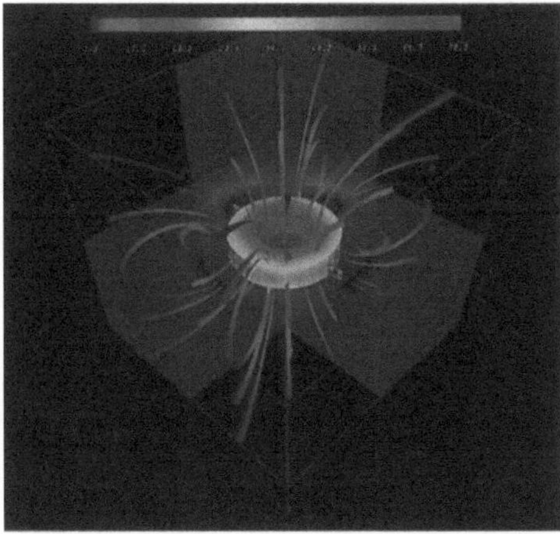

Fig. 12. The complete magnet, stream ribbons, with part of the surrounding air

3.6 Using colour

Some experiments were made with colour, including using a reversed hue ramp
with red for low magnitude and blue for high - this gave some eye-catching visualizations since shades of red dominated the image. Unfortunately this wasn't
what the users of the electromagnetic package expected and so the more traditional blue to red hue ramp was used. The **harwell colormap** was used to
perform gamma correction.

3.7 Using Stereo

Although stereo displays would help many viewers resolve depth, the most realistic views for this dataset were obtained with the Cambridge Autostereo Display [7]. This is a research display that does not require the use of special eye glasses to a produce 3D effect for grey-scale images. The viewer's head can move from side to side and still experience a realistic effect. Unlike normal stereo displays which require 2 views, this system uses 8 or 16 views to produce a single static picture. Small animations were also possible.

4   Visualizing Transient solutions

The Vector Fields post-processor, OPERA, views transient solutions by displaying the real and imaginary phases as two separate, static views. By computing additional steps in the cycle it is possible to explore the time-varying flow with AVS.

The images so far were from the same point in the cycle. When the objects from the three circular streamline probes (shown in Figure 9) were mirrored, although a lot of field lines resulted, it was still possible to distinguish them, see Figure 13. Using this many seed points for transient solutions does not always give such good results.

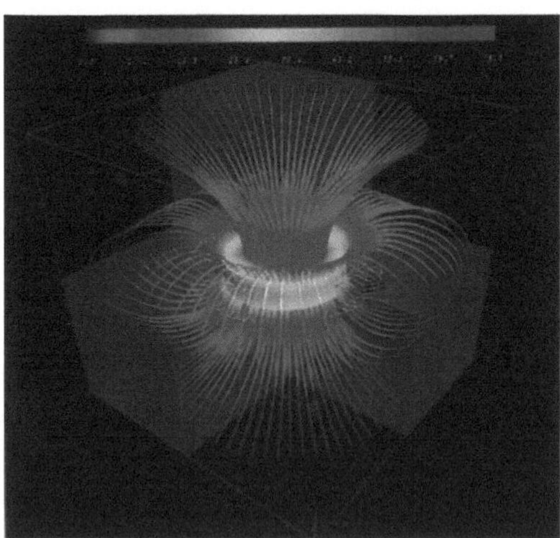

Fig. 13. Two sets of Stream ribbons and one set of streamlines from circular probes, mirrored

Direct manipulation to select viewpoint and to choose specific solutions proved invaluable to explore single solutions, but to observe the changing flow through time, the more systematic approach of stepping through the cycle from beginning to end at regular intervals gave better results. Use of the AVS Animator allowed key frames to be defined to determine an animation. For viewing static solutions, selecting key frames with varying camera positions could give smooth tumbling effects. For transient data, the easiest way to comprehend the changing flow patterns was to keep the viewpoint static. Rather than move the viewpoint during the animation, it was found less confusing to stop the animation, reposition the camera to resolve any visual ambiguities (e.g. by rotation) and when satisfied restart the animation cycle.

It was not possible, on the workstation used, to compute the transient solutions and produce visualizations in real-time. For simple wire-frame/tube renderings each frame took a few seconds, but more complex images took approximately 5 minutes each. This made it difficult to follow changes but was acceptable when trying to find unusual flow patterns. To overcome the slow response, a sequence of frames were saved as images then played back in real-time. AVS could not play back the images fast enough so, ImageMagick, a public domain collection of image processing tools, was used instead. Use of a more powerful, supercomputer could also overcome these speed problems.

The frames, in Figure 14, give an idea of the cyclic flow. The same three circular probes used in Figures 9 and 13 are used to generate the streamlines. Note since the real vectors are much larger than the imaginary, not much changes in the early parts of the animation, i.e. between $0°$ and $80°$. At $90°$ the field lines appear to be collapsing around the magnet.

Some positions, of the probe caused the AVS streamline module to crash. e.g. at $275°$, which should match $95°$ with the direction of the flow reversed. When the cycle from Figure 14 was continued, strange flow lines appeared just above $90°$, see Figure 15.

By trying to use as many seed points as possible it was also difficult to determine which probe(s) produced this effect.

Unlike the patterns observed, the flow lines were expected to form loops (possibly clipped by the boundary of the air). By going back to the cones representation from Figure 6, it was discovered that the flow was so small from $90°$ to $95°$ and $270°$ to $275°$ near the coil that trying to interpolate streamlines/ribbons was not sensible.

The three circular probes were carefully positioned at the beginning of the cycle to give distinguishable sets of field lines, but by $90°$ this was no longer the case. As the vectors away from the magnet were too small to give accurate field lines, a more reasonable region to investigate around this time, was close to, or on, the magnetic material - this would give an idea of residual fields when the current no longer flowed. In Figure 16 the field lines, from the circular probe near the inner edge of the magnet, at $90°$ are shown as ribbons. The magnet was now rendered to show material type. With the material identifier rendered it was possible to see how the material type effected the magnetic field.

In Figure 17 the field lines, from the circular probe near the outer edge of the magnet, are shown at $94°$ and $95°$. The direction of flow is also reinforced with cones. Initially the cones were coloured grey to avoid confusion with the hues

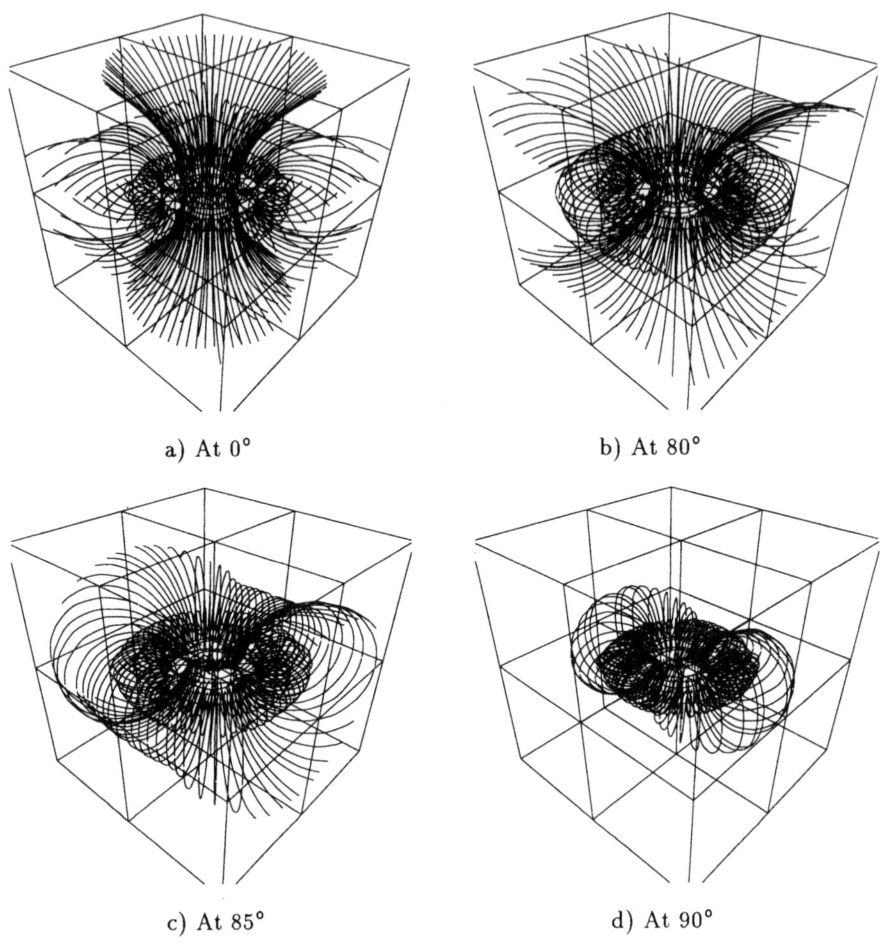

a) At 0° b) At 80°

c) At 85° d) At 90°

Fig. 14. 4 frames from the cyclic animation

used to represent data, but this did not transfer from the screen to photographic media. Instead a less saturated red was used for the outside and similarly reduced saturation yellow for the inside cones. Since the direction of the flow is preserved through the magnet, the cones were mirrored in the upper half only. If the cones were mirrored in the lower half the direction would be opposite to reality.

The field lines no longer form loops around 95° at the outer edge of the magnet but not until 102° near the inner edge. These findings do correlate in with the underlying physics of the problem. Since the driving current is zero at 90° and at 270°, it is not surprising the vectors tend to zero away from the magnet. In addition, above 90° and 270° the direction of the current is reversed and so, for a number of degrees above these two points in the cycle, small vectors can continue to become smaller. Also some delay can be expected between the change in current and its effects on the cells furthest away in the magnet, and so the vectors on the outer edge tend to zero before the inner edge.

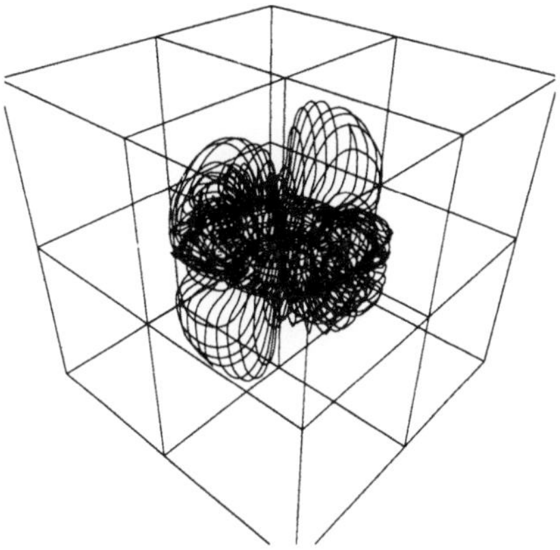

a) At 92° 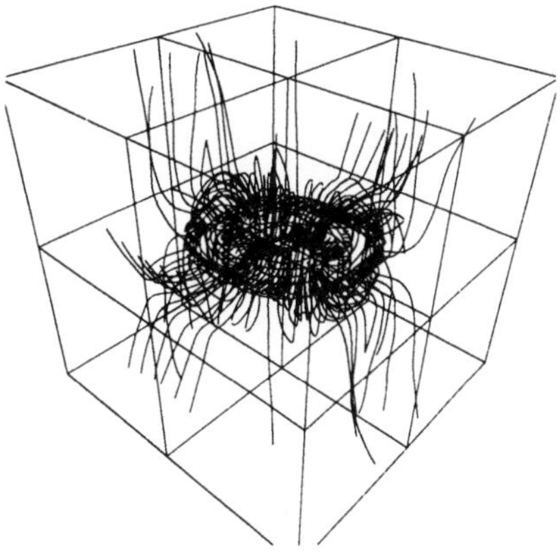 b) At 93°

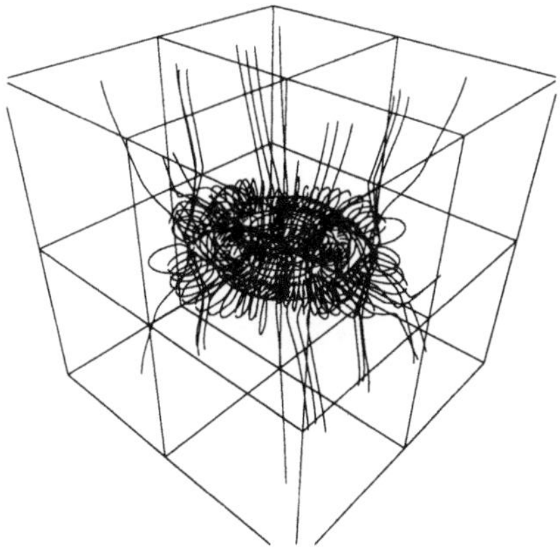

c) At 94° 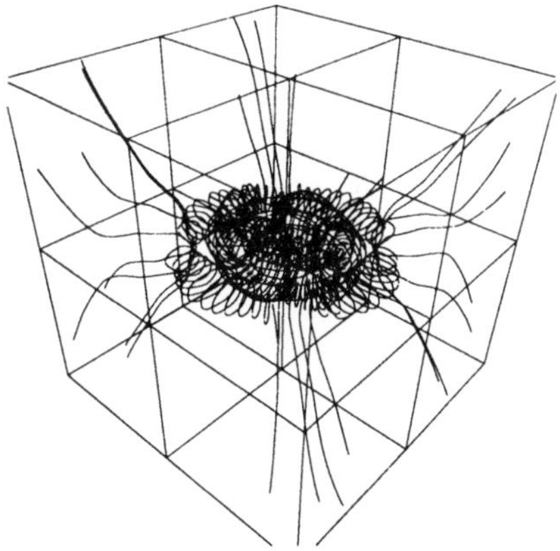 d) At 95°

Fig. 15. 4 more frames - concentrating on the residual fields in the magnet

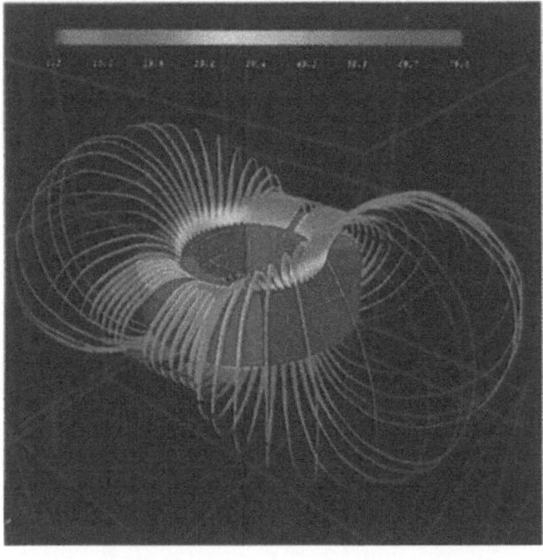

Fig. 16. Magnet shaded by material type with field lines as ribbons at 90°

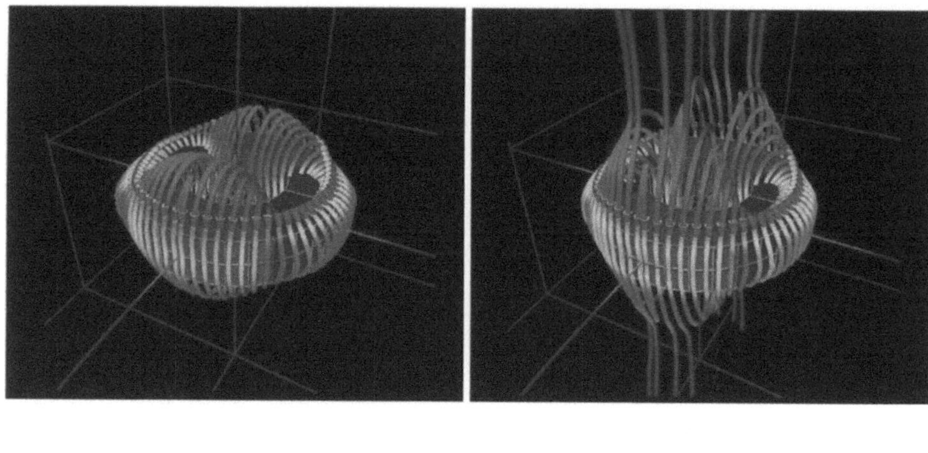

a) At 94°   b) At 95°

Fig. 17. Magnet shaded by material type with field lines as tubes at 94° and 95°

5 Discussion

Visualizing time-varying 3D vector field data is a non-trivial task. Using a visualization package to process electromagnetic data proved a quick and effective way to produce interactive applications. They give access to facilities not normally associated with Finite Element specific packages, for example to experiment with colourmaps and to produce smooth animations. New tools could be added but, without access to vital source code, modifications to existing modules were not possible. In addition, some routines in the AVS libraries are not generally available to the average user, so re-writing standard modules with modifications or additional features is not always feasible.

The main drawback in using a visualization system is the considerable demands on computing and memory resources. Unless a supercomputer is available, this can prove limiting for large datasets. Even the relatively small solution used for this paper pushed the SG Crimson to it's limits with the more complex AVS networks. There are packages which can process unstructured flow data much more efficiently, for example the binary to VISUAL3 [8] is available in the public domain for several platforms. Unlike AVS, these do not have the flexibility to allow the user to add new facilities and do not provide general purpose utilities like the Animator.

Direct manipulation to change the viewpoint and to re-orient probes was found the best method for exploring the data. When real-time response was not possible it was found better to produce animations, then playback in real time. Sometimes, e.g. when trying to find interesting streamlines in the cycle, some delay between frames was acceptable. Changing viewpoint while cycling through the time varying data caused confusion, since it was then difficult to concentrate on the changing field lines.

To explore transient solutions, a more systematic approach proved more successful. Stepping through the cyclic flow at 5° intervals highlighted the regions to concentrate on.

It is important to try several techniques to understand complex flow. Switching between visualization techniques, as from streamlines to cones, can clarify what is really going on.

Good communication between the source of the data, i.e. the scientist or engineer, with the visualization expert is essential. Some experimentation can be made by the visualization user, but without a good understanding of the underlying science these are purely 'pretty pictures'. It was possible, with the applications developed, to discover unusual flow patterns during the cycle and then by alternating visualization techniques to decide this was probably due to trying to interpolate streamlines in regions where too many vectors approached zero rather than any real phenomena. This was confirmed by looking at the underlying physics of the problem.

Some areas of flow visualization can rely on feature extraction to find regions of interest, e.g. field topology [9]. Although these features do not occur in electromagnetics it is hoped that future work for more general flow problems will incorporate this method for unstructured grid data.

References

1. AVS: The AVS 5.0 User's Guide. Advanced Visual Systems Inc., 300 Waltham, Ma 02154, USA.
2. AVS: The AVS 5.0 Module Reference manual. Advanced Visual Systems Inc., 300 Waltham, Ma 02154, USA.
3. AVS: The AVS 5.0 Developer's Guide manual. Advanced Visual Systems Inc., 300 Waltham, Ma 02154, USA.
4. Vector Fields: ELEKTRA/OPERA Reference Manual. Vector Fields Ltd., 24 Bankside, Kidlington, Oxfordshire, UK.
5. Larkin, S. (ed): Evaluation of Visualization Systems. AGOCG Technical Report 9. May 1992 (UK).
6. Post, F.H., van Wijk, J.J: Visual Representations of Vector Fields - Recent Developments and Research Directions. ONR Workshop on Data Visualization, Darmstadt, Germany, July 1993. (to be published by Academic Press, Inc)
7. Dodgson, N.A., Lang, S.R: Visualization on the Cambridge Autostereo Display (3DTV). UK Visualization Community Club, Proceedings 6, "3D Visualization in Engineering Research". 1 day seminar held at RAL, 24th March 1993 (UK).
8. Giles, M.B., Haimes, R: Advanced Interactive Visualization for CFD. Computing Systems in Engineering, Vol. 1 No. 1, pages 51-62. 1990.
9. Helman, J.L., Hesselink, L: Visualizing Vector Field Topology in Fluid Flows. IEEE Computer Graphics and Applications, Vol. 11 No.3. May 1991.

Possibilities and Limits in Visualizing Large Amounts of Multidimensional Data

Daniel A. Keim, Hans-Peter Kriegel

Institute for Computer Science, University of Munich
Leopoldstr. 11B, D-80802 Munich, Germany
{keim, kriegel}@informatik.uni-muenchen.de

Abstract. In this paper, we describe our concepts to visualize very large amounts of multidimensional data. Our visualization technique which has been developed to support querying of large scientific databases is designed to visualize as many data items as possible on current display devices. Even if we are able to use each pixel of the display device to visualize one data item, the number of data items that can be visualized is quite limited. Therefore, in our system we introduce reference points (or regions) in multidimensional space and consider only those data items which are 'close' to the reference point. The data items are arranged according to their distance from the reference point. Multiple windows are used for the different dimensions of the data with the distance of each of the dimensions from the reference point (or region) being represented by color. In exploring the database, the reference point (or region) may be changed interactively, allowing different portions of the database to be visualized. To visualize larger portions of the database, sequences of visualizations may be generated automatically by moving the reference point along some path in multidimensional space. Besides describing our visualization technique and several alternatives, we discuss some of the perceptual issues that arise in connection with our visualization technique.

Keywords: Visualizing Multidimensional Multivariate Data, Visualizing Large Data Sets, Perception of Visualizations

1. Introduction

The progress made in hardware technology allows today's computer systems to store very large amounts of data. The available storage space is easily filled with data that is often automatically recorded via sensors and monitoring systems. Today, even simple transactions of every day life, such as paying by credit card or using the telephone, are typically recorded by using computers. Even larger amounts of data are generated by automated test series in physics, chemistry or medicine and satellite observation systems are expected to collect one terabyte of data every day in the near future [FPM 91]. Usually, many parameters are recorded resulting in multidimensional data with a high dimensionality. The data of all areas mentioned so far is collected because people believe that it is a potential source of valuable information providing a competitive advantage (at some point). Finding the valuable information hidden in them, however, is a difficult task. With today's database systems and its query tools, it is only possible to view quite small portions of the data. If the data is presented textually, the amount of data that can be displayed is in the range of some one hundred data items, but this is like a drop in the ocean when dealing with data sets containing millions of data items. Having no possibility to adequately query and view the large amounts of data that have been collected because of their potential usefulness, the data becomes useless and the database becomes a data 'dump'.

For the exploration of very large amounts of multidimensional data to be successful in the near future, we believe that it is essential to make the human being an integral part of the data analysis process. It will be important to combine the best features of humans and computers. The intelligence, creativity and perceptual abilities of humans which are unmatchable need to be supported by computers which are best suited to do searching and number crunching. Some five years ago, a broader community of researchers recognized the potentials of visualization techniques to analyze and explore large amounts of data. With visualization techniques, larger amounts of data can be presented on the screen at the same time, colors allow the users to instantly recognize similarities or differences of thousands of data items, the data items may be arranged to express some relationship and so on. Over the last years, many techniques for the visualization of multidimensional data have been developed. It seems, however, that many of the techniques do not provide adequate support for the flood of data we are facing today. Since, on the other hand, the technology for generating, collecting and storing data is available, the gap between the amount of multidimensional data that should be visualized and the amount of data that can be visualized is growing. Additionally, in most systems the perceptual abilities of humans are only used to a very limited extend; only few systems use e.g. motion and sound to help the user in data analysis. Therefore, a major research challenge is to find human-oriented ways to help the user in exploring large amounts of multidimensional data.

In this paper, we focus on our visualization technique that uses color and dense displays to visualize multidimensional data. Our visualization technique (see section 2 for a brief description) has originally been developed in the context of querying large databases, but it has proven to be more generally useful for visualizing large amounts of data with an arbitrary dimensionality. In section 3, some extensions, display alternatives and other ideas will be presented. In section 4, we then provide examples that show the possibilities and limits of our visualization technique.

2. The Basic Idea of our Visualization Technique

Visualization of data which have some inherent two- or three-dimensional semantics has been done even before computers could be used for visualization, and since using computers for this purpose, a lot of interesting visualization techniques have been developed by researchers working in the graphics field. Visualization of large amounts of arbitrary multidimensional data, however, is a relatively new research area. Researchers in the graphics/visualization area are currently exploring techniques in different application domains. Examples are shape coding [Bed 90], worlds within worlds [FB 90], parallel coordinates [ID 90], iconic displays [PG 88, BMS 92], dimensional stacking [LWW 90], hierarchical plotting [MGTS 90] or dynamic methods as presented in [MZ 92]. In most of the approaches proposed so far, the number of data items that can be visualized on the screen at the same time is quite limited (in the range of 100 to 1000 data items), but it is a declared goal to push this limit [Tre 92]. In dealing with databases consisting of millions or even billions of data items, our goal is to visualize as many data items as possible at the same time to give the user some kind of overview of the data. The obvious limit for any kind of visual representation is the resolution of current displays which is in the order of one to three million pixels, e.g. in case of our 19 inch displays with a resolution of 1024 x 1280 pixels there are about 1.3 million pixels. Our idea is to use each pixel of the screen to give the users visual feedback about the data, allowing them to easily focus on the desired data and to understand the influence of multiple parameters.

The basic idea of our visualization technique for large data sets is described in [KKS 93]. In dealing with databases consisting of billions of data items with multiple dimensions (often ten and more parameters), we had to find an adequate way of restricting the amount of data to be visualized to a number that can be displayed on the screen. In our approach, for this purpose reference points (or regions) in multidimensional space are introduced and only the data items that are 'closest' to the reference point are visualized. The 'closeness' is determined using distance functions for each of the dimensions. The distance functions are datatype and application dependant and must be provided by the application. Examples for distance functions are the numerical difference (for metric types), distance matrices (for ordinal and nominal types), lexicographical, character-wise, substring, phonetic or semantic difference (for strings) and so on. In the specification of the reference region, not all of the dimensions have to be used. If m of the n dimensions are used in the specification of the reference point, then the reference region itself is an (n-m)-dimensional space with some extension into the other m dimensions. Dimensions that are not used in the specification of the reference region have basically no impact on the visualization since the distance for such dimensions is zero for all data items.

Having calculated the distances for each of the dimensions which are part of the reference point specification, the distances are combined into the closeness factor. Important aspects such as normalizing and weighting the distances of the different dimensions, the formulas used to calculate the closeness factors and the heuristics used to reduce the number of displayed data items are described in [KKS 93]. The closeness factors are then sorted resulting in a one-dimensional distribution ranking the data items according to their closeness. The basic idea for visualizing the data items is to map the value ranges of the different dimensions to color and represent each data item by multiple pixels being colored according to the distance values for each of its dimen-

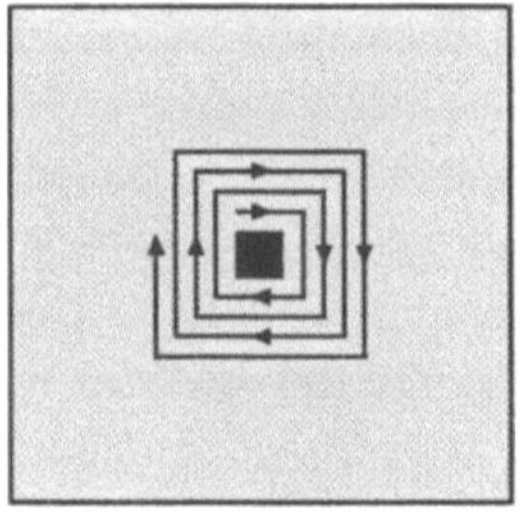

Figure 1: Spiral Shaped Arrangement of the Data Items

sions. To maximize the number of just noticeable differences, we use a colormap with constant saturation, an increasing value (intensity) and a hue (color) ranging from yellow over green, blue and red to almost black. The colormap is continuous except for a discontinuity between yellow and green which is used to distinguish the data items inside the reference region from those outside the reference region. The colored pixels are then displayed on the screen with data items fitting into the reference region centered in the middle of the window and the other data items are arranged rectangular spiral shaped around this region (c.f. figure 1) according to the overall closeness factor. A separate window is provided for each of the dimensions. In these separate windows, the pixels for each data item are placed at the same relative position, allowing the user to relate the visualization of the different dimensions. In figures 5-10, several visualizations of four- and six-dimensional data are provided. The data sets used to generate the visualizations are artificially generated data sets with explicitly inserted multidimensional clusters. A detailed description of the examples will be given in section 4.

After getting the visual feedback, the user may interactively change the reference point (or region). Using highlighting of corresponding pixels in different windows or a projection of the visual representation to specific color ranges, the users may further explore the data helping them to relate the distances for the different dimensions. By having the possibility to get the attribute values corresponding to some specific color, the users may better understand and interpret the visualizations. According to the discoveries made during this process, the user may then incrementally change the reference point (or region) using sliders provided for each of the dimensions. For details about the interactive interface see [KKS 93].

3. Alternative Visualization Techniques

In this section we describe some extensions, alternative visualization techniques and additional ideas, all being related to our main idea for visualizing large amounts of multidimensional data that has been described in section 2.

Alternative 1: Mapping two Dimensions to the Axes

An idea for an alternative screen layout is to display the data in 2D with selected attributes assigned to the axis. The problem with conventional 2D or 3D representa-

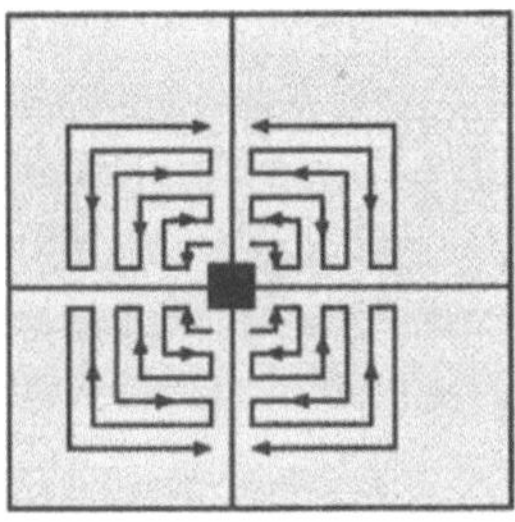

Figure 2: 2D-arrangement of the Data Items

tions is that on the one hand many data items may be concentrated in some area of the screen while other areas are virtually empty, and on the other hand many data items are superposed and therefore not visible. Although conventional 2D or 3D visualizations may be very helpful, e.g. in cases where the data have some inherent two- or three-dimensional semantics, we did not pursue this idea for several reasons: One reason is that in most cases the number of data items and dimensions that can be represented on the screen at the same time is quite limited. This was in contrast to one of our goals, namely to visualize as many data items as possible on the screen. A second reason is that in most cases where a 2D or 3D arrangement of the data is straightforward, systems using such arrangements have already been built.

Stimulated by the conventional 2D or 3D data representations, we got the idea for a second kind of visualization which includes some feedback on the direction of the distance for distance functions that provide positive and negative distance values. The basic idea is to assign two attributes to the axis and to arrange the distances according to the direction of the distance; for one attribute negative distances are arranged to the left, positive ones to the right and for the other attribute negative distances are arranged to the bottom, positive ones to the top. Inside the regions, the data items with the closeness factors sorted in an descending order are arranged from the middle (yellow region) to the edges of the window (see figure 2). With this kind of representation, we do not represent the distance of data items directly by its locations, but we denote the absolute value of the distance by its color and the direction with respect to the dimensions assigned to the axes by its location relative to the correct answers. An advantage is that each data item is assigned to one pixel and that data items with the same distance are not superposed. A problem may occur in some special cases if e.g. no data items exist that have a negative distance for both attributes but many data items that have a negative distance for one of them and a positive one for the other one. In this case, the bottom left corner of the window would be completely empty. In the worst case, two diagonally opposite corners of the window may be completely empty (c.f. figure 8) and, as a result, only half as many data items as possible are presented to the user. Even in this case, the user gets valuable information on how to change the reference point (or region) in order to get the desired results. In general, we found that maximizing the number of visualized data items conflicts with arrangements that directly visualize distances by different locations on the screen.

An open questions is which of the dimensions should be assigned to the axes. Since not only the dimensions that are used in the specification of the reference region, but all dimensions may be used as axes dimensions, the number of choices may be quite high. If we deal with n-dimensional data and all of the dimensions have positive and negative distances, we have $\sum_{i=1}^{n-1} i = \frac{n \times (n-1)}{2}$ possibilities to choose two of them to be assigned to the axes. This means that for 5-dimensional data, there are already 10 possibilities and for 15-dimensional data there are 105 possibilities. For data sets with a high dimensionality, it is not practicable to try all combinations. If the user has no preferences for the axes dimensions, the system needs to support the user in selecting them. One possibility would be to automatically generate a sequence of visualizations, presenting the data set with all possible assignments of dimensions to the axes. According to the visual impression from the sequence, the user may then decide which of the assignments are interesting and useful for data exploration. Further research will be necessary to examine the impact of assigning different dimensions to the axes and to find criterions for choosing the right combination of dimensions to be assigned to the axes. In figures 6-8, we provide some example visualizations, comparing different assignments of dimensions to the axes. We also compare the original and the 2D-visualization technique, showing some of their advantages and disadvantages. The details about the example visualizations are described in section 4.

Alternative 2: Grouping the Dimensions for each Data Item

In both, the original arrangement and the 2D arrangement just presented, the pixels corresponding to the different dimensions of the same data item are distributed in the different windows for each of the dimensions. Another visualization alternative is to present all dimensions for one data item grouped together in one area. The areas each representing one data item may be arranged rectangular spiral-shaped according to the closeness factor of the considered data items (see figure 3). The coloring of the distances for the different dimensions may be the same as in the original or 2D arrangement. The generated visualizations, however, will be completely different than the ones of the original and 2D arrangement since they consist of only one window with many areas visualizing all dimensions of the considered data items instead of many windows each providing a visual representation of only one dimension of the considered data items. At this point, it should be mentioned that the idea of grouping the dimensions into one area is similar to the shape coding approach described in [Bed 90]. In our approach, however, we do not focus on the shape to distinguish the data items and also the criterion and kind of arranging the data items is different.

First experiments show that for the grouping arrangement more pixels per data value are needed. According to our experience, at least 4-times (better 9 or 16-times)

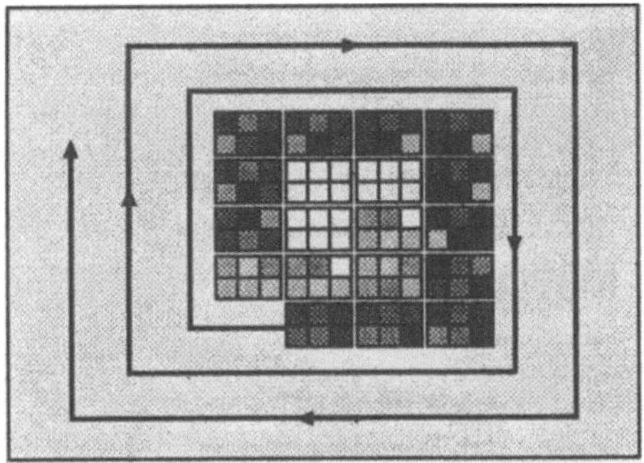

Figure 3: Grouping the Dimensions

as many pixels are needed per data value when compared with the other arrangements. This means that only one-fourth (one-ninth or one-sixteenth) of the data items can be displayed on the screen at one point of time. Note, that additional pixels are needed for surrounding the area for each data item. In contrast to the other arrangements, a border is necessary; otherwise it would be impossible to know which pixels belong to one data item. In figure 5, an example data set with 2000 data items is visualized using the original, the 2D-, and the grouping technique. The original and the 2D-arrangement are enlarged by 100%, whereas the visualization of the grouping arrangement is reduced to about 70% of its original size.

Despite the fact that only fewer data items may be visualized, we expect the grouping arrangement to provide more useful visualizations for data sets with larger dimensionality. In the original and 2D arrangement, the pixels for each dimension of the data items are only related by their position. For relatively small dimensionality (e.g. less than 8 dimensions), it seems to be quite easy for humans to relate the different portions of the screen. The larger the dimensionality gets, however, the more difficult is it to relate the different parts of the visualization and to perceive correlations across them. In case of the grouping arrangement it is not necessary for the user to relate different portions of the screen and therefore, for larger dimensionality the arrangement may be advantageous.

Alternative 3: Time Series of Visualizations

In trying to visualize larger amounts of data than possible with the techniques described so far, an important potential is to consider time as an additional dimension. For many applications it is natural to consider time sequences of visualizations describing some features which are changing over time. In the terminology of our system, this could be described as moving the reference point (or region) along the time dimension. Most traditional systems for visualizing time series consider in each step only the data items at a certain point of time. Contrarily, with our visualization technique we consider all data items that are 'close' to the reference point (or region) including data items with differing time values as long as their overall closeness factor with respect to the reference point (or region) is high enough.

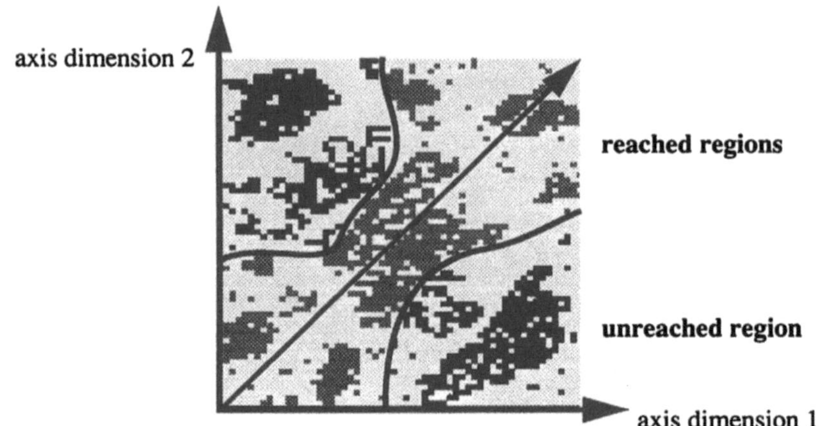

Figure 4: Reached and Unreached Region in Moving the Reference Point in 2D

Our idea for visualizing larger portions of the database is to generalize the technique of generating sequences of visualizations by moving the reference point (or region). Instead of moving the reference point (or region) along the time axis, the user may choose an arbitrary path through n-dimensional space. Obviously, the semantics of the derived visualizations are different. If moving the reference point (or region) along some parameter, e.g. the temperature in an environmental database, the user may get insight in the corresponding distributions of the other parameters such as ozone or CO_2. The user may also choose more complicated paths through n-dimensional space, e.g. by varying two parameters such as temperature and ozone at a time. The specification of more complicated paths, however, is not straightforward and it is not clear how the user will be able to deal with the complicated semantics introduced by complex paths. An open question is which paths provide visualizations that are 'easy' to perceive and allow the user to find the interesting data sets. In figure 4, we show an example for the reached and unreached regions when moving the reference point diagonally in a two-dimensional data set. Since the percentage of data items that is displayed at one point of time is constant, the portion of two-dimensional space that is reached is not a regular section parallel to the diagonal. An interesting question is how it may be guaranteed that the whole database (or a given portion of it) is covered. Future work is necessary to answer this question and to find intuitive ways in dealing with path specification and the semantics of the resulting visualizations.

Despite the unsolved problems, in searching a very large database of multidimensional data for interesting correlations, clusters or hot spots, our technique seems to be a promising approach since neither the number of data items that can be visualized nor their dimensionality is limited and the visualizations may help the user to get important, previously unknown information out of the automatically generated visualization sequences.

4. Evaluating the Usefulness of our Visualization Technique

In this section, we describe our first experiences in evaluating our visualization techniques. We will give some examples for visualizations of multidimensional data and we will discuss some open questions which we believe to be important for future research. Most of the presented issues do not only apply to our technique but also to visualization techniques developed by other researchers. Our goal in presenting the questions is to stimulate the discussion about current visualization techniques for large amounts of multidimensional data. The data used to produce the presented visualizations are artificial data sets with specific characteristics. In evaluating different visualization techniques, the possibility to precisely control the characteristics of the test data (e.g. the correlation coefficient of two dimensions, the distribution function of some of the dimensions, the location, size and shape of clusters, etc.) is crucial. The details of the program used to generate the test data sets are beyond the scope of this paper. A general discussion of test data sets for evaluating data visualization techniques can be found in [BKP 94].

In the following, we provide some examples that illustrate the possibilities and limitations of our visualization techniques. In figure 5, we present a generated data set with 2000 six-dimensional data items using all three visualization techniques. The original and 2D-arrangement are enlarged, whereas the grouping arrangement is reduced in size. While some clustering is visible in all three visualizations, the clustering is most obvious in the 2D arrangement. In comparing the three visualization techniques, we found that the grouping arrangement provides useful visualizations for rather small data sets (100 - 1000 data items), while the original and 2D-technique work for much larger data sets (up to 100.000 data items and more).

In figures 6-8, we compare the original and the 2D technique. One first observation is that in many cases the 2D-visualizations will show more of the structure than the visualizations generated using the original technique. In figure 6 for example, no structure is visible in the original arrangement (left part of figure 6) while the corresponding 2D-arrangements (right part of figure 6 and both parts of figure 7) clearly show multiple clusters. The 2D-visualizations in figures 6 and 7 only differ in the choice of the axes dimensions. Note, that different clusters are not visualized equally well with different axes assignments; in some cases, clusters may even not be visible at all. In comparing the original and the 2D-visualization technique, we found that each of them has some advantages and disadvantages. A clear advantage of the 2D-technique is that it provides more information than the original arrangement (c.f. figure 6). A disadvantage, however, is that the number of data items that can be visualized is lower. Figure 8 shows an example for a 2D-visualization which has two opposite quadrants that are completely empty.

Figure 9 presents two visualizations of the same data set that only slightly differ in the percentage of data items that are presented. In the left part, 100% of the 10.000 data items are presented while in the right part only 95% of the data items are displayed. The data set used to generate the visualizations of figure 9 contains a few data items for each dimension that have a much higher value than the remaining data items. Since the data values are normalized after reducing the number of data items to the desired percentage, the coloring of the visualizations in the right part is much better than in the left part. Note, that the factor by which the high values are higher than the remaining data items is different for each of the dimensions. For dimension one the factor is 1, for dimension three the factor is 2, for dimensions four, five and six the factor is 4, and for dimension two the factor is 6.

In figure 10, we present two visualizations showing 5-dimensional clusters in 6-dimensional data. The data used to generate the visualizations consists of 17000 data items. Two-third of the data is generated randomly (in the range [0, 100] for each of the dimensions) and the remaining one-third of the data defines three five-dimensional clusters. The three clusters have been inserted at well-defined locations of the 6-dimensional space. The only difference between the left and right visualization in figure 10 is that the clusters are at different locations. As reference region, in both cases we used the 6-dimensional rectangle with [0, 10] for each of the dimensions. Interesting is that in the windows for the sixth dimension some additional clustering appears. We also experimented with four-dimensional clusters in six-dimensional space. We found that they are not perceivable at all. Although we expected lower dimensional clusters to be less perceivable, it was surprising that the perception was diminishing that fast with a smaller dimensionality of the cluster. By adapting the weighting factors, however, we found a way to make the 4-dimensional clusters perceivable. If the weighting factors on the cluster dimensions are significantly higher than on the other dimensions, then a cluster with lower dimensionality will be perceivable. Changing the weighting factors implies a change of the shape of the multidimensional region around the reference point (or region) which contains the data items that will be visualized. It also induces a change in the ordering of the data items and will therefore result in completely different visualizations.

We further found that, in most cases, the extension of the cluster in multidimensional space has only a minor effect on the visualization. More important is the percentage of data items that form the cluster. Small clusters are only perceivable if they are close to the reference point and have distinctly different characteristics than the remaining data items. The percentage of data items that need to be part of the cluster for the cluster to be perceivable depends on the distinctness between base data and cluster, on the dimensionality of the data, and on the cluster's distance to the reference point. The latter problem can be resolved, for example, by inverting the ordering of data items in the visualizations which causes data items with larger distances to be closer to the center and therefore to be more visible.

Interesting topics for future research are an examination of the type of information (type of clusters, type of correlations, etc.) that is perceivable with our visualization techniques, an examination of the impact of different weighting and distance functions, and a comparison of the different visualization techniques for multidimensional data that have been proposed so far. One important step towards an in-depth examination of current visualization techniques for multidimensional data will be an integrated test data generation and evaluation tool which is currently being implemented at our institute. The tool will allow to generate artificial data sets with given characteristics. The test data sets may be described by the distribution functions for each of the dimensions, the correlations or functional dependencies between the dimensions, and the clusters which again may have arbitrary characteristics. The tool may be used to evaluate one single visualization technique to find its strength and weaknesses but it will also be helpful to compare different visualization techniques to find out which technique is most suitable for which types of data.

5. Summary and Conclusions

Visualizing very large amounts of arbitrary multidimensional data is one of the big challenges that researchers in the graphics/visualization area are currently facing. The task is to efficiently find interesting data sets, i.e. hot spots, clusters of similar data or correlations between different parameters. In this paper, we briefly presented our approach for visualizing large amounts of multidimensional data. It allows to visually represent the largest amount of data that can be displayed at one point of time on current display technology. Alternative visualization techniques and additional features have been described. Many questions that arise in connection with the perception of our visualization techniques have been brought up focussing on some of the possibilities and limitations of visualizing large amounts of multidimensional data. In trying to find the answers for these questions, the goal of our future research is improve the perception of our visualization techniques and to find new ways of pushing their limits to be able to visualize even larger amounts of data with an even higher dimensionality.

References

[Bed 90]   Beddow J.: *'Shape Coding of Multidimensional Data on a Microcomputer Display'*, Visualization '90, San Francisco, CA, 1990, pp. 238-246.

[BKP 94]   Bergeron R. D., Keim D. A., Pickett R.: *'Test Data Sets For Evaluating Data Visualization Techniques'*, in: Proc. Workshop on Perceptual Issues in Visualization, Springer, 1994.

[BMS 92]   Bergeron R. D., Meeker L. D., Sparr T. M.: *'A Visualization-Based Model for a Scientific Database System'*, in: Focus on Scientific Visualization, eds: Hagen H., Müller M., Nielson G., Springer, 1992, pp. 103-121.

[FB 90]   Feiner S., Beshers C.: *'Visualizing n-Dimensional Virtual Worlds with n-Vision'*, Computer Graphics, Vol. 24, No. 2, 1990, pp. 37-38.

[FPM 91]   Frawley W. J., Piatetsky-Shapiro G., Matheus C. J.: *'Knowledge Discovery in Databases: An Overview'*, in: Knowledge Discovery in Databases, AAAI Press, Menlo Park, CA, 1991.

[ID 90]   Inselberg A., Dimsdale B.: *'Parallel Coordinates: A Tool for Visualizing Multi-Dimensional Geometry'*, Visualization '90, San Francisco, CA, 1990, pp. 361-370.

[KKS 93]   Keim D. A., Kriegel H.-P., Seidl T.: *'Visual Feedback in Querying Large Databases'*, Visualization '93, San Jose, CA, 1993, pp. 158-165.

[LWW 90]   LeBlanc J., Ward M. O., Wittels N.: *'Exploring N-Dimensional Databases'*, Visualization '90, San Francisco, CA, 1990, pp. 230-239.

[MGTS 90]   Mihalisin T., Gawlinski E., Timlin J., Schwendler J.: *'Visualizing Scalar Field on an N-dimensional Lattice'*, Visualization '90, San Francisco, CA, 1990, pp. 255-262.

[MZ 92]   Marchak F., Zulager D.: *'The Effectiveness of Dynamic Graphics in Revealing Structure in Multivariate Data'*, Behavior, Research Methods, Instruments and Computers, Vol. 24, No. 2, 1992, pp. 253-257.

[PG 88]   Pickett R.M., Grinstein G.G.: *'Iconographic Displays for Visualizing Multidimensional Data'*, Proc. IEEE Conf. on Systems, Man and Cybernetics, Beijing and Shenyang, China, 1988.

[Tre 92]   Treinish L. A., Butler D. M., Senay H., Grinstein G. G., Bryson S. T.: *'Grand Challenge Problems in Visualization Software'*, Visualization '92, Boston, Mass., 1992, pp. 366-371.

Figure 5: Example visualizations generated using our three visualization techniques

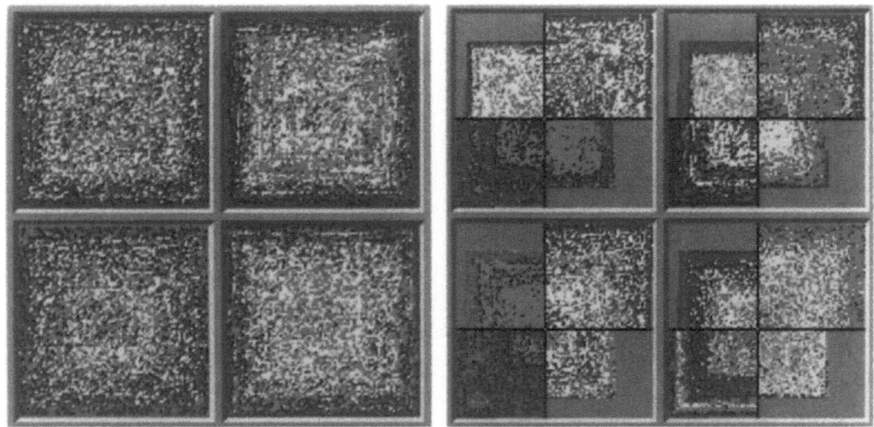

Figure 6: Advantage of the 2D visualization technique

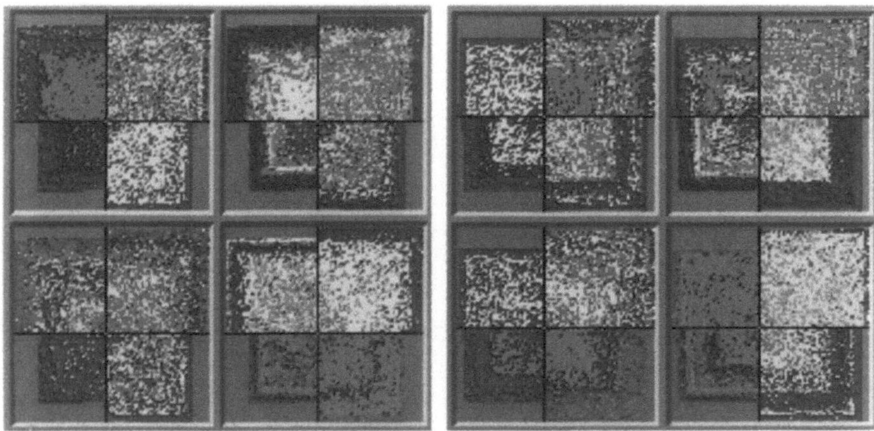

Figure 7: Effect of assigning different dimensions to the axes

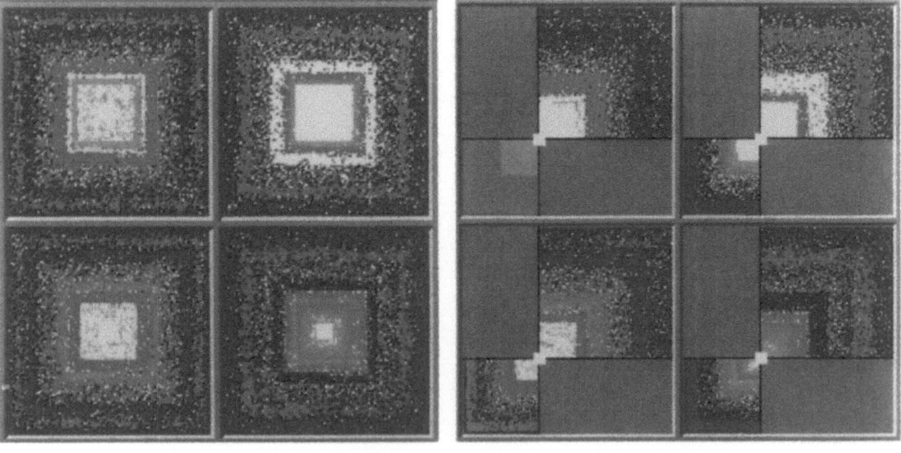

Figure 8: Disadvantage of the 2D visualization technique

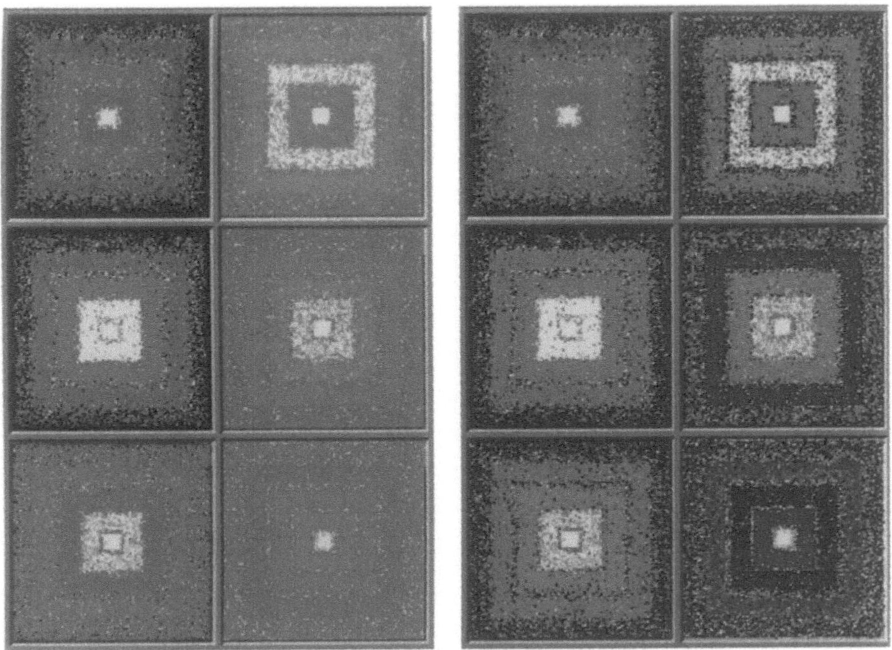

Figure 9: Effect of reducing the amount of data by 5%

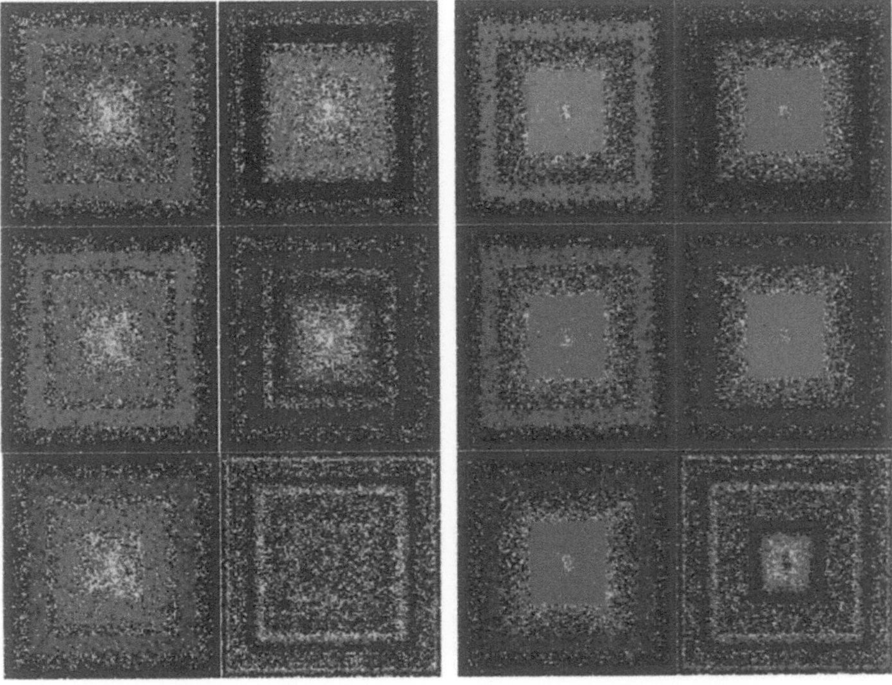

Figure 10: Visualization of 5-dim. clusters in 6-dim. data

Using a Lattice for Visual Analysis of Categorical Data

G. Deon Oosthuizen*

Department of Computer Science, Stanford University

Stanford, CA94305, USA

Fritz J. Venter

Department of Computer Science, University of Pretoria

Pretoria, 0001, South Africa

Abstract

This paper addresses the problem of visualizing data of dimension higher than three. The method described is primarily aimed at so-called categorical (or nominal) data, but also applies to other discrete-valued data, i.e. data entities defined in terms of vectors of discrete attribute values. Categorical data is data with unordered attribute values, mostly because they are symbolic. Data with boolean values constitute a special case.

The method described utilizes the fact that data entities can be regarded as points in n-dimensional space, where n is the number of attributes recorded for each data entity. Although it is not possible to depict an n-dimensional cube using three dimensional graphics, it is possible to provide the user with an abstraction of the structure of the data by revealing the non-empty subspaces and how they are related. The graphical user interface allows the user to browse through the space of abstractions.

1   Introduction

Most current work on visualization is aimed at the analysis of numeric and, more specifically, real-valued data. However, some databases contain data that is inherently non-numeric, e.g. when attributes of entities are colours, shapes, formats, arrangements, etc. Apart from the fact that a graphical plot of individual variables is not meaningful - the attribute values cannot be *ordered* along an axis - the mapping between variables involves discontinuous functions which do not lend themselves to two- or three-dimensional axis-based displays. This paper complements the other papers in this volume by addressing the problem of visualizing discrete-valued data of dimension higher than three.

The main objective of data analysis is to determine if there is any hidden regularity in data that is not visible to the human eye. The existence of such regularity normally reflects dependencies between the variables or attributes recorded, i.e. it reflects structure in the data. We present a method which provides the user with a complete inventory of regularities in a discrete-valued data set. By imposing an n-dimensional arrangement on the data (see below),

---

the user is enabled to inspect them systematically and thereby gain insight in the phenomenon under investigation.

The method described utilizes the fact that data entities can be regarded as points in n-dimensional space, where n is the number of attributes recorded for each data entity. Although it is not possible to depict the n-dimensional cube or *hypercube* involved using three dimensional graphics, it is possible to provide the user with an abstraction of the structure of the data by revealing the non-empty subspaces and how they are related.

We will describe how the model provides the functionality:

- to select a set of data points [1], on the basis of an arbitrary specification of attribute values, and obtain a complete visual profile of its structure - i.e. to visualize the subsets (clusters) and their identifying and distinguishing attributes.

- to select any two attributes and display visually their intricate logic and probabilistic relationships.

- to select any two sets of attributes and obtain a visual display of their relationship as well as a visual account of how additional (or fewer) attributes in each set would affect the relationship.

The framework enables the user to visually explore the structure of any subclass of data points by specifying an abstract class description consisting of one or more attributes. The system displays an exhaustive visual breakdown of the *structure* and *relationships* of the specified set - a unique property provided by the lattice.

We first define the lattices used in Section 2. Then we explain how they are displayed (Section 3) and used (Section 4) to visualize structure and dependencies in data. In Section 5 we briefly comment on the computational complexity of the task of generating the lattices, and in Section 6 on related work.

2   Concept Lattices

A lattice is a directed acyclic graph in which every pair of nodes have a unique nearest common descendant - or *meet* - and a unique nearest common ancestor - their *join*. The lattices discussed here are of a special kind, called *concept lattices* [11], which have the following additional properties:
- Apart from the children of the universal node at the top and the parents of the NULL node at the bottom no other nodes in the graph have exactly one parent or exactly one child.
- No node has a parent (i.e. no node is directly linked to another node) to which it is also indirectly linked by means of a path that goes via one or more other nodes.

Although Wille formally introduced the notion of a concept lattice [11], the particular organization of data described here also corresponds to the so-called *cladistic* approach to classification used by biologists and linguists for some time [3]. In fact, the fundamental idea behind concept lattices dates back

---

[1] I.e. a data entity consisting of an array of attribute values.

to Aristotle who noted the inverse relation between the number of properties required to define a concept and the number of entities to which the concept applied. This is referred to as *the duality of intension and extension* [10].

In the practical implementation to be described in this paper the universal node - which ensures that all pairs have joins - and the *null* node - which ensures that all pairs of nodes have meets - are omitted. Consequently, every pair of nodes in the current lattices has a unique meet or no meet, as well as a unique join or no join. This does effect the other properties of the lattices, though.

A concept lattice is constructed by creating a node for each data point at the bottom of the graph (e.g. nodes E1-E9 at the bottom of fig. 1 represent the corresponding data records in Table I) and a node for each attribute-value at the top. During this process internal nodes are created between the data points at the bottom and the attributes at the top (they are marked with *'s in fig. 1). Each data point is then connected to its respective attributes whilst ensuring that the graph remains a lattice (see fig. 1). It can be shown that a given set of entities give rise to a unique lattice. The exact manner in which the lattice is constructed is beyond the scope of this paper (algorithms can be found in [1], [2] and [7]).

Each internal node denotes a pattern of attributes that occurred in more than one data point. Since all possible combinations of attributes could potentially occur in the data, the potential number of nodes in a lattice is equal to the size of the powerset of the number of attributes per entity, i.e. 2^n, where n is the number of attributes per entity. The actual number of nodes realizing in real world data sets is, however, only a fraction of this amount (see Section 5). This is also the reason why the lattice is useful as a data analysis tool: each of the internal nodes created represents a regularity in the data. And to ensure that the graph is not cluttered by accidental coincidences, statistically insignificant nodes are removed from the lattice (see Section 5).

Let us consider an example. Fig. 1 shows a lattice generated from the data shown in Table 1. (For the sake of simplicity, some nodes have been omitted.) It depicts underground sample information regarding rock samples collected for laboratory analysis.

Entity no.	Size	Colour	Shape	Contains Heavy Metals	Structure
E1	small	brown	regular	yes	hard
E2	large	brown	irregular	no	brittle
E3	large	yellow	regular	yes	hard
E4	small	black	regular	no	brittle
E5	large	black	regular	yes	brittle
E6	large	brown	regular	yes	hard
E7	large	black	irregular	no	brittle
E8	small	brown	irregular	no	brittle
E9	large	brown	irregular	yes	brittle

Table I.

The lattice in fig. 1 appears rather irregular, but this is because we are representing an n-dimensional structure in two-dimensional space, and only the patterns of attributes that actually occurred in the data are represented. Fig. 2 shows the same lattice but preserves more of the regularity in two dimensions and gives a taste of the visualization capabilities afforded by the operations described in subsequent sections.

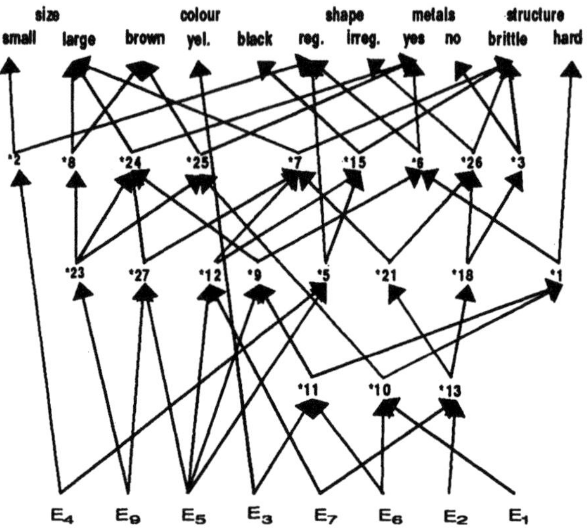

Fig. 1

By abstracting a subspace, the user can be presented with a regular structure (see fig. 4). I.e. the higher dimensional spaces is visualized by making two-dimensional cuts through the structure.

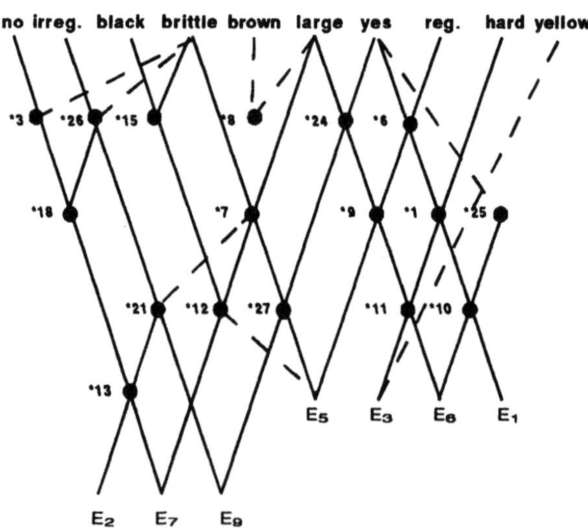

Fig. 2

While the user is traversing the lattice, one specific node is regarded as the focus of interest and only its relationships to its neighbours are considered, as we will explain below.

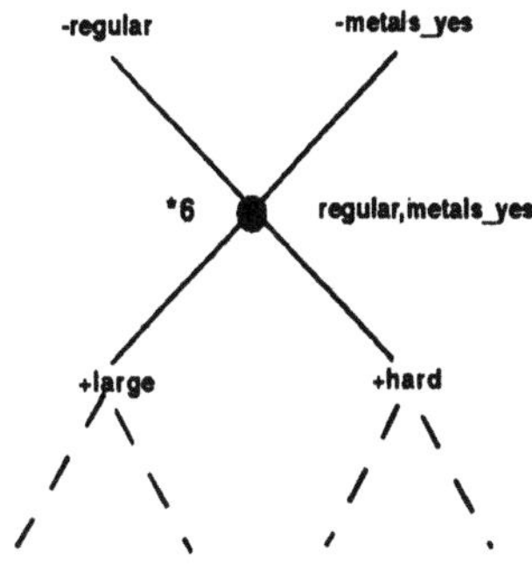

Fig. 3.

3   Visual display of the lattice

Unlike the toy example above, real lattices are too large and complex to display on the screen in their entirety. Consequently only the relevant parts of the lattice are displayed (see fig. 3).

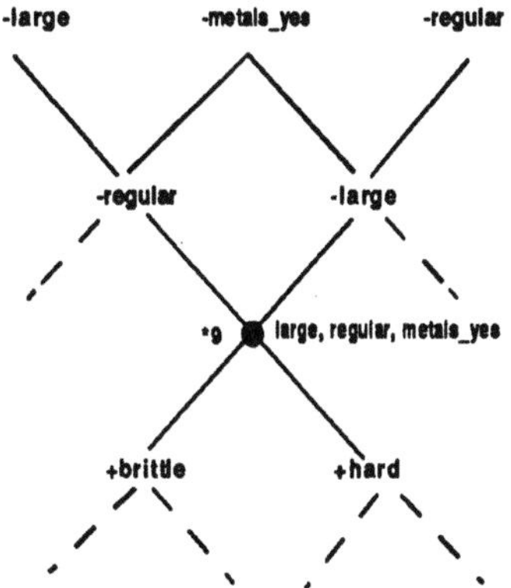

Fig. 4.

A popup window is used to display a list of all the recorded attributes. It is displayed initially when the user selects attributes to be investigated, and is removed thereafter, and displayed only if the user wants to select attributes from the list which are not present in the visible part of the lattice.

The lattice is represented as nodes labelled by attribute names, linked by lines. The attribute names are preceded by a + or a -, as we will explain in the next section. At every stage one node is defined as the *focus* and highlighted and displayed in the middle of the screen. Differently from the other nodes, the focus is (temporarily) labelled with all the attributes the user selected, until a new focus is selected. If the user clicks on a node, it is moved to the center of the screen, and becomes the new focus. The cardinality of the set of data points covered by each node in the graph is referred to as the node's *strength* and can optionally be displayed at the node. (We refer to the nodes below a given node as the nodes *covered* by it and the nodes above it as the nodes *spanned* by it.)

If node *6 of fig. 2 is the current focus, the screen appears as shown in fig. 3. If the user now selects node *9, the screen is updated as shown in fig. 4.

4   Visualizing the data

4.1   Visualizing clusters

The internal nodes represent groups of entities that are *similar* in the sense that they share a particular set of attributes. Thus, each of the internal nodes represents a *cluster*, and the lattice provides us with an *n*-dimensional clustering (or *clumping*, the technical term for clusters that overlap). The lattice has the unique capability of providing an exhaustive visual account of all the possible hierarchies of clusters that can be formed starting with the different attributes as roots. Also, the hierarchies are optimally overlapped.

Data analysis proceeds as follows. The user specifies a category of data that s/he is interested in. This is done by selecting one or more attributes from the popup window. If s/he selects a single attribute, a portion of the lattice is displayed such that the node in the lattice corresponding to the chosen attribute is placed in the middle of the screen and highlighted.

If more than one attribute are selected, their meet is computed and a portion of the lattice is displayed such that the meet is placed in the middle of the screen. The description of the class of data points selected can thus be either *simple*, consisting of one attribute, or *composite*, consisting of several attributes. We refer to the centred node as the current *focus*.

Firstly, the screen display provides the user with an exhaustive breakdown of the dataset selected. It is as if the user zoomed in on a subset of the data and is now able to view at a microscopic level the threads that run between the entities studied. The hierarchy below the focus provides him with the following information.

For example, the data-set characterized by (the attributes above) the focus *6 consists of two major subclusters, namely *9 and *1 (see fig. 2). The points in *9 all have the additional property **large** and the points in *1 have the distinguishing property **hard** (see fig. 1 and fig. 3). *9 and *1 are not necessarily mutually exclusive. In other words, the subclusters may overlap (see *11 in fig. 2). By selecting *9, the user shifts the focus down to *9. The screen is updated immediately to reflect the composition of the data, i.e. to show the attributes that distinguish the clusters in the vicinity of the focus from it. The new focus is labelled with the full description of the cluster it represents. With *9 as focus the user is shown what major subclasses *9 is composed of (see fig. 4). Each of the classes represented by nodes directly above the focus differs from it by *not* having exactly one of its attributes.

Just as biological hierarchies of insects or plants provide scientists with understanding of how subspecies are related, the user is here able to run up and down the hierarchy, getting an overall picture of the composition of the data. The difference is that the current hierarchies are *induced from raw data* and provides insight into structure yet *unknown* to the user. Each dimension (each attribute) provides a different angle for viewing the data - each attribute-node is the root of a different hierarchy. The hierarchies do overlap, but are optimally integrated, i.e. no single subhierarchy is repeated at more than one place in the lattice. Thus the lattice provides an incredibly well consolidated view of the data. At a glance the user can observe all the hierarchies one particular cluster forms part of. We are not aware of any statistical analysis tools which have similar visualization capabilities.

Secondly, the lattice displays possible causes for the description displayed at the focus (the attributes characterizing the focus). Groups of two or more '+' node labels below the focus could potentially be the cause of the set of attributes contained in the description of the focus. This will be further discussed when the relationships between attributes are addressed in the next section.

Notice that the n-dimensional visual ordering provided by the lattice is the *ultimate* ordering, given the n attributes. There is no 'better' or more refined ordering of the data possible, unless more information is provided. On the other hand, any lower-dimensional ordering or coarser-grained arrangement would not be able to utilize and reflect the total content of the information available. The lattice enables the user to visualize the distribution of the data points in multidimensional space, by focusing on an abstraction (a subspace) and extending or reducing it on different 'sides' and by observing the density of data points involved.

4.2   Visualizing dependencies between attributes

In the previous section we considered the nodes around the focus, and specifically the subclusters below it. The attributes spanned by the focus were merely displayed as a list at the focus, without depicting any further structure. However, the precise composition of the tree-like structure stretching upward from the focus, with the focus as its root and the attributes as leaves constitutes a rich source of information regarding the dependencies between the attributes in the list [2].

These dependencies are computed as follows:

1. Find the meet of two or more given attributes (we call it the set Q).

2. Compute the upward closure [3] of the meet.

3. If the upward closure contains any attributes apart from those in Q, then these attributes are implied by the ones in Q.

The meet of the set Q is the root of a tree of which the leaves are the elements of Q. E.g. if Q = {**large**, **irregular**}, then the meet of Q is *21, which spans **brittle** in addition to the members of Q. By definition the meet is *unique* and *maximal*: keeping in mind that *21 is the set of data points that has the attributes **large** and **irregular** in common, it means that *21 is the *only* set of data points with these properties, and it includes *all* the data points which have the attributes. But, the class of data points covered by *21 are characterized by the attributes spanned by it. In other words, *all* the points concerned have *all* the attributes. We may conclude, therefore, that

---

[2] This is true for all internal nodes, of course. Not only for the focus.

[3] We refer to the set of nodes above a given node as its *upward closure*, and to the nodes below it as its *downward closure*.

150

for all x in the given data set such that
x has attributes large and irregular, x also has attribute brittle.

I.e. **brittle** can be *inferred* from the other two properties (A formal proof can be found in [8]).

On the other hand, **large** and **brittle** does not imply **irregular**. This is because the meet of **large** and **brittle** is *7 and not *21, and *7 does not span any additional attributes. The reader may verify this in Table I.

Thus, each node in the lattice signifies dependencies between the attributes it spans. The potential number of dependencies that could exist is equal to number of ways in which the attributes spanned by a particular node can be grouped together - each possible subset of the attributes may or may not determine one or more of the rest of the attributes. From this total number of dependencies the lattice structure neatly excludes (inhibits) all invalid inferences. This is accomplished by a meet spanning the given attributes *exclusively* - as in the case above where *7 inhibits inferring **irregular** form **large** and **brittle** by spanning them exclusively.

It is important to note that, by definition, the meet delineates uniquely and comprehensively all dependencies between the attributes it spans. There is no chance of an additional relationship between the same nodes, hidden somewhere else in the lattice.

Obviously, the validity of the statements regarding dependencies we made above is limited to the data set observed. A small number of data points like the ones in the example do not justify generalization but in real world data sets the situation may be different, i.e. the data may be regarded as representative of the whole domain.

4.2.1 *Implementation*

Initially the user selects a number of attributes. These are then displayed with the focus as before, but unlike in Section 4.1, the inferred attributes are not listed with the given ones, but in a second popup window. Thus the selected attributes are displayed at the focus, and the inferred attributes in a popup window. Previously, **brittle** would be listed with the focus, i.e. the full description of the cluster *21 would be displayed at the node . Here it is kept separately in the popup window to enable the user to distinguish between *given* and *inferred* attributes. By selecting nodes below or above the focus in the lattice, attributes may be added to or removed (resp.) from the popup window.

4.2.2 *Causal Analysis*

In the above example, one attribute was inferred, but it is possible that multiple attributes could be inferred. As a result it is possible for a user to visually inspect whether one set of attributes depends on another. Different sets of attributes (A_i) may be selected and sets of attributes (B_i) would be inferred.

In the opposite scenario where a set B of attributes is selected and the set A of attributes it depends on is required, the labels below the focus have to be considered again. Typically, this functionality is needed during causal analysis. Different combinations of attributes below the focus may be selected in order to determine which combinations imply the attributes of the focus, i.e. which combinations could potentially cause the attributes of the focus [4].

Of significance is the fact that it is not a *list* that is displayed. A complete visual profile is provided. The reason why this is important, is that the potential number of combinations of attributes that could form causes are mostly too large to inspect one by one. The visual display of the layered topology of the lattice structure enables the user to comprehend the interplay between the attributes involved in order to conduct an informed heuristic search through the possible combinations.

4.2.3 Rules

Whereas in Section 4.1 we were concerned with the recognition of dominant clusters and the identification of their descriptions, we are here concerned with the relationships between attributes. In essence, the focus displays the left hand side, and the popup window the right hand side of a *production rule* (as found in contemporary Expert Systems), or rather a tentative [5] production rule. However, instead of just deriving a host of rules and displaying them on the screen, the lattice allows one to navigate through the 'rule-space' in a manner which enables one to notice and get a feeling of the relationships between sets of attributes (or *concepts*) denoted by nodes in the graph - an aspect that is totally lost in the 'flat' structure of a rule base. (The associated description of each node 'implies' the description of each node above it.) The nearest (or most similar) rule in each dimension can be detected immediately.

4.3   Visualizing probabilistic relationships

The ways of visualizing the data described in the previous two sections were based on the logical relationships between attributes, as derived from the data. Statistical noise often blurs a clear view of structure in data. Therefore it is important to be able to visualize probabilistic relationships between attributes.

Visualization of probabilistic relations is supported by the lattice as follows. The cardinality of the subclasses *9 and *1 of the class *6 in fig. 2 are

$$| *9 | = 3 \text{ and } | *1 | = 3$$

where $| *6 |$ denotes the cardinality of class *6 as reflected by the strength of the node in the graph. Then

$$| *6 | = | *9 | + | *1 | - | *11 | = 4.$$

We will not give a formal motivation here, but informally this means that since three out of the four data points in class *6 had attribute **large**, the conditional probability of **large** given the attributes of *6 is .75. (Again, the number in the toy example are too small to warrant the conclusion).

Attributes spanned by nodes below the focus, but not by the focus itself, are inferred with varying degrees of uncertainty, all of which can be read from the graph. These probabilistic relationships may be visualized firstly by noting the positions of the '+'-labelled nodes (e.g. see fig. 4) relative to that of the focus. Attributes lower down are inferred with a smaller probability (by the selected

---

[4] Whether or not a combination of attributes is a potential cause of the focus depends on whether the attributes in question have a meet above the focus or not.

[5] If the number of nodes covered by the focus (i.e. the meet) is small, one might regard the implicated dependencies with caution. However, more data may confirm or contradict it.

attributes). Additionally, the link lengths can be employed to portray the relative strengths of nodes, and hence the probabilistic relationships between the attributes appearing in their respective labels.

This would enable the user to track down attributes that are suggested by or closely associated with the selected attributes.

5 Computational complexity

The potential of the lattice size to grow exponentially with the number of attributes per data point and the computational complexity involved in constructing such a large structure is a matter of concern. This problem can be alleviated by compressing the lattice. The compressed lattice is obtained by removing internal nodes from the bottom of the lattice and by connecting their children (the data points) to their parents in a straightforward way, disregarding lattice properties. In other words, by constructing a hybrid structure which is half lattice, half directed graph, we can maintain a graph containing an arbitrary number of nodes [9].

The computational cost of constructing a lattice for typical datasets is $O(n^3)$, where n is the number of attributes per entity. However, this cost is off-line and not involved during data exploration described above.

6 Visualizing numeric data

A premise to the discussed use of lattices in order to visualize data dependencies is that the domain data is represented as tuples of discrete attribute-value assertions, i.e. each data sample is described in terms of a finite set of symbols.

However, if the domain of discourse includes numeric variables, then they have to be categorized into a finite set of intervals which represent all possible attributes per variable. For example, let us assume we want to determine and visualize the dependencies between p_i, where $i \in [1,N]$. For each p_i we define ranges, say 'high','med' and 'low'. The set of attributes are then:

$p_1$high,$p_1$med,$p_1$low,$p_2$high,$p_2$med,$p_2$low ... $p_N$high,$p_N$med,$p_N$low

The value of p_i is then evaluated to determine in which interval ($p_i$high,$p_i$med or $p_i$low) it falls. This interval will then be the element of the N-tuple (sample) in position i.

This method of transforming numeric data into symbolic interval descriptions is actually a generalization of the method used in qualitative physics where all numeric values are classified as *positive* or *negative*. Although the above example uses three intervals per parameter, any number of intervals can be used per parameter and each parameter need not have the same number of intervals. To determine these intervals, it is suggested that a statistical pre-processing phase should be used to determine frequency distributions of each parameter. The input data to this phase will thus be the actual numeric values of the parameters. The ranges (on the x-axis of each parameter's histogram) forthcoming from this process can then be used to guide the calculation of ranges to associate with each symbolic interval name of a parameter (e.g 'high','med','low'). The use of frequency distributions ensures that the symbolic intervals of the parameters are well representative of the data.

It should be noted that this mechanism for visualization of numeric parameter dependencies allows scalability of the degree of detail of the dependencies between parameters which is depicted. On the one side of the scale, i.e. when we have only 2 intervals per parameter, we have effectively generated another view of what qualitative physics would produce. On the other end of the scale, when the number of intervals per parameter is very large, we have a much more detailed view of the dependencies between the parameters. The maximum number of intervals to be used per parameter can be indicated by the number of intervals which have representation larger than a certain threshold with respect to the histograms mentioned above. Another phenomenon which can be used to guide the choice of the number of intervals per parameter, is the growth of the resulting lattice as a function of the number of intervals per parameter. When the number of intervals per parameter is increased, the number of values, which fall within each interval (the surface under the frequency distribution) will decrease. This and the increased number of attributes will result in more nodes forming in the lattice. A threshold in terms of the potential number of nodes in the resulting lattice can then be used as guideline to determine the maximum number of intervals per parameter.

7 Other applications and related work

Concept lattices have been applied to the fields of Machine Learning - the internal nodes represent generalizations [8], and Machine Translation - to index transfer rules. Its application to object management in class repositories is described in [6]. Lattices of sizes up to twenty four thousand nodes have been constructed. Data sets with up to thirty five attributes have been processed. The time to insert a data point depends on the size of the lattice. For twenty-attribute data points it ranges from half a second for a lattice of a hundred nodes to one minute for a lattice with six thousand nodes on a DEC 5000/240 work station. However, this affects the (off-line) construction of the lattice which happens once only. The interactive operations described in this paper, which involve node location and retrieval, take only in the order of a second. The lattice construction algorithm is written in C. It employs so-called marker propagation and is amenable to parallel implementation [5]. Most experimental work was done on PCs, however.

Lattices have been used in knowledge processing (knowledge representation and truth maintenance) before [12], but with two exceptions [1][2] as virtual structures only. The actual structures were not constructed, mainly because of knowledge of their worst case exponential growth. We explained how that problem can be counteracted.

8 Conclusion

Much work on data exploration and visualization has been done in the area of analysing real-valued data. Numerical methods are often forced onto nominal data by assigning numerical values to categories in order to be able to perform statistical analysis, which is mostly based on calculating distances between vectors. This imposes an ordering on the data-values which may lead to incorrect analysis [4]. Also, numerical calculation of distances is normally based on the *overall* differences between points: if two points are very similar regarding most of their attributes but differ substantially with respect to one or two attributes, they may appear very dissimilar, although they are in fact quite similar.

The above problems are eliminated by the method described in this paper. The lattice provides a novel way of visualizing structure in discrete-valued data by enabling the user to home in on a desired section of the attribute space or of the entity space. The total ordering imposed on the data allows the user to alter his view incrementally by proceeding in tiny steps from one subspace to another, guided by signposts at varying distances revealing what will be encountered if the focus is shifted in a particular direction. This is possible because of the unique capability of a lattice of joining extension (clusters and their elements) and intension (the description/meaning of each class) in one structure.

Finally, the user may perform a unique kind of what-if analysis on the attributes in order to gain insight in complex relationships between them. Disjunctive class descriptions were not discussed, but could be catered for by means of a special operation.

9   Acknowledgements

We are indebted to Nils Nilsson for providing us with a stimulating environment, to the Foundation for Research Development for supporting the research financially, and to our Creator.

References

[1] C. Carpineto, G. Romano: GALOIS: An order-theoretic approach to conceptual clustering. Proceedings of the International Machine Learning Conference, Amherst, 1993, pp.33-40.

[2] R. Godin, R. Missauoui, A. Hassan: Learning Algorithms using a Galois Lattice Structure. Proceedings of 1991 IEEE International Conference on Tools for AI, San Jose, pp.22-29.

[3] H.M. Hoenigswald, L.F. Wiener: Biological Metaphor and Cladistic Classification: An Interdisciplinary Perspective. University of Pensylvania Press, Philadelphia, 1987.

[4] N. Jardine, R. Sibson: Mathematical Taxonomy. John Wiley & Sons, 1971.

[5] G.D. Oosthuizen: A Production System Based on Massively Parallel Marker Propagation. In D.J. Evans, G.R. Joubert, F.J. Peters (Eds.), Parallel Computing 89. Elsevier Science Publishers (North Holland), 1990.

[6] G.D. Oosthuizen, C. Bekker, C. Avenant: Managing Classes in Very Large Class Repositories, Proceedings of TOOLS-2 Conference: Technology of Object-Oriented Languages and Systems, Paris, 1990, pp.625-633.

[7] G.D. Oosthuizen: Lattice-Based Knowledge Discovery. In Proceedings of AAAI-91 Knowledge Discovery in Databases Workshop, Anaheim, 1991, pp. 221-235. (AAAI=American Association for Artificial Intelligence, Menlo Park, CA)

[8] G.D. Oosthuizen, D.R. McGregor: Induction through Knowledge Base Normalisation. In Proceedings of ECAI-88, Pitman Publishing, 1988, pp. 396-401.

[9] G.D. Oosthuizen: The application of Concept Lattices to Machine Learning. Submitted for publication.

[10] J.F. Sowa: Conceptual Structures: Information Processing in Mind and Machine. Addison-Wesley, 1984.

[11] R. Wille: Restructuring Lattice Theory: An Approach Based on Hierarchies of Concepts. In I.Rival (Ed.) Ordered Sets. Reidel, Dordrecht - Boston, 1982, p. 445-470.

[12] W.A. Woods: Understanding Subsumption and Taxonomy: A Framework for Progress. In J.F. Sowa (Ed.) Principles of Semantic Networks. Morgan Kaufmann Publishers, 1991, pp. 45-93.

Audience Dependence of Meteorological Data Visualization

Florian Schröder

Fraunhofer Institute for Computer Graphics
Wilhelminenstr. 7, 64283 Darmstadt, Germany

Abstract: The decisions how to visualize a specific data set depend mostly on the kind of data itself. But in order to visualize the data in a perceptually effective way, the audience who will view the results must also be considered. The importance of audience dependence varies between the different fields of applications of visualization techniques. The visualization of meteorological data is a field where audience dependence plays a very important role in determining the way of turning the data into images. We have developed a system for visualizing weather-related data for meteorological researchers as well as for lay TV audience. Both groups have very specific demands on the results. Meteorologists need a presentation containing their symbols and possibly many data sets at the same time to get a better understanding of their data and simulation models. Lay audience need images or sequences they can understand intuitively and easily with their every day experience of weather phenomena. We categorized the types of meteorological data and determined the demands of the two groups. To visualize cloud-specific weather data for the lay audience, e.g., we incorporated fractal functions to show clouds that look like real clouds. The result so far is our system TRITON II, which is used for meteorological research and for the daily production of forecast videos for several German TV stations.

1   Audience Dependence

Data cannot be visualized in a perceptually effective way, when the audience is not taken into account. When scientists need to visualize data gathered from experiments or simulations, several aspects should be considered to find the "right" way to do so. First, the visualization methods are mostly determined by the type of data itself. Furthermore, the medium of presentation (e.g., paper, film, or monitor screen) plays an important role. In addition, the decisions on interpolation types, mapping or coloring, e.g., depend sometimes heavily on the audience who will view the results of the visualization process: Will the final images be presented to colleague researchers who are familiar with the data and their context, or will these images be viewed by other scientists with less information on background or context? Are the pictures of the gathered data prepared for the members of an executive management, for prospective funding sources, or for a non-scientist audience, e.g., via newspapers, journals, television, or other media? Scientists want to see results that closely relate to the original data, whereas viewers from decision-making authorities expect "polished" images that present progress or new features. Finally, an audience with no scientific background knowledge of the data needs a simple presentation, a visualization result they can understand easily, fast, and intuitively. This latter kind of visualization is the subject of this article. Although every kind of data requires specific processing and presentation techniques, and the specific needs of the audience must always be taken into account, there are some cases where the audience dependence is especially important: For example, there are historical data, medical science, environmental data, computational fluid dynamics, oceanography, finite element simulations, cartography (geographic information systems), or archeological data, just to name a few. And there is one field, where the viewers' demands clearly dominate: meteorological data in weather forecasting presentations!

Weather observation has always been a fascinating topic. Today, this is even more the case with forecasting getting increasingly precise. Farmers need to know the weather development in advance for professional reasons; government authorities want precise weather information to limit damages in catastrophic events, such as extremely high tides or hurricanes; military decisions rely on it for strategic reasons, and airports for safety reasons. But also the television audience is very interested in weather forecasting. Sure evidence for this is that there are TV stations broadcasting only weather forecasts, or the fact that TV stations receive a lot of mail when they change the style of their weather forecasting presentation.

2   Visualizing Meteorological Data for Non-Experts

When meteorological data must be visualized for non-expert audience, the presentation form and the applied visualization methods are very different from the standard cases:

2.1   Ground Image

The ground image over which the meteorological data will be visualized is extremely important. It provides the geographical context of the meteorological data. It makes no sense to show clouds, e.g., over a uniform background that gives no hint about the location of the presented weather phenomena or its scale.

Especially in a television weather forecast, the ground image must allow quick and easy orientation. The main information that shall be presented is the weather development in the future. But before the viewer can focus his or her interest on a certain region (e.g., home town or location of vacation resort), it must be spotted in the television image. Only then can the viewer concentrate on the actual weather development in that area. And since the forecast is usually presented very briefly, the orientation must be quick and easy.

There are several possibilities of generating a ground image from geographical information. One possible way is to map satellite photographs as a texture onto the terrain elevation data, in order to get a realistic model of the ground. This solution was rejected by us, since average unexperienced viewers have great difficulties in recognizing presented areas, especially when unique landmarks like large mountain chains or coast lines are not present in the image. This is because the colors in such photographs are then determined by the vegetation or by man-made landmarks, and viewers will then start to search for airports or big cities. But this can take too much time in a short television weather forecast presentation.

The method of coloring the ground with its elevation data according to the colors used in school atlases results in a much quicker orientation. Most people are very familiar with seeing atlases or globe models and with finding known regions merely based on the height color patterns that are characteristic for that region. By adding rivers, borders, lakes, and locations with the names of cities, the orientation can easily be even further improved, of course.

Since the ground image only provides the geographical context to the relevant prognosis data, the colors used here must not be too intense and distracting to the eye of the viewer who should focus on the relevant information.

Fig. 1. A ground image allowing quick and easy orientation

2.2   Fractal Clouds

To present simulated clouds to a lay audience, it is essential to find a visualization method that uses cloud objects. The goal is to find a presentation where the viewer can fully concentrate on comprehending the weather development rather than spending too many thoughts on understanding the presentation itself.

The currently used European Model (at the Deutscher Wetterdienst, DWD) for simulating the weather development over continental Europe has a grid resolution of about 50 km x 50 km. This results in often only a dozen times a dozen data entries for the prediction area. With this extremely coarse resolution it is impossible to generate realistic-looking clouds with conventional visualization techniques.

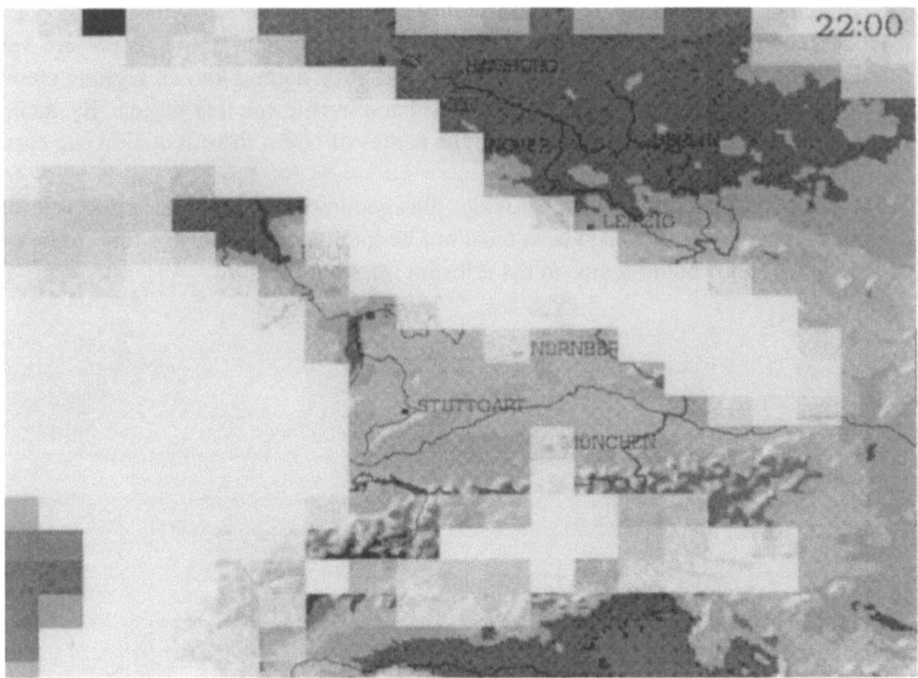

Fig. 2. Cloud-specific simulation data on original grid

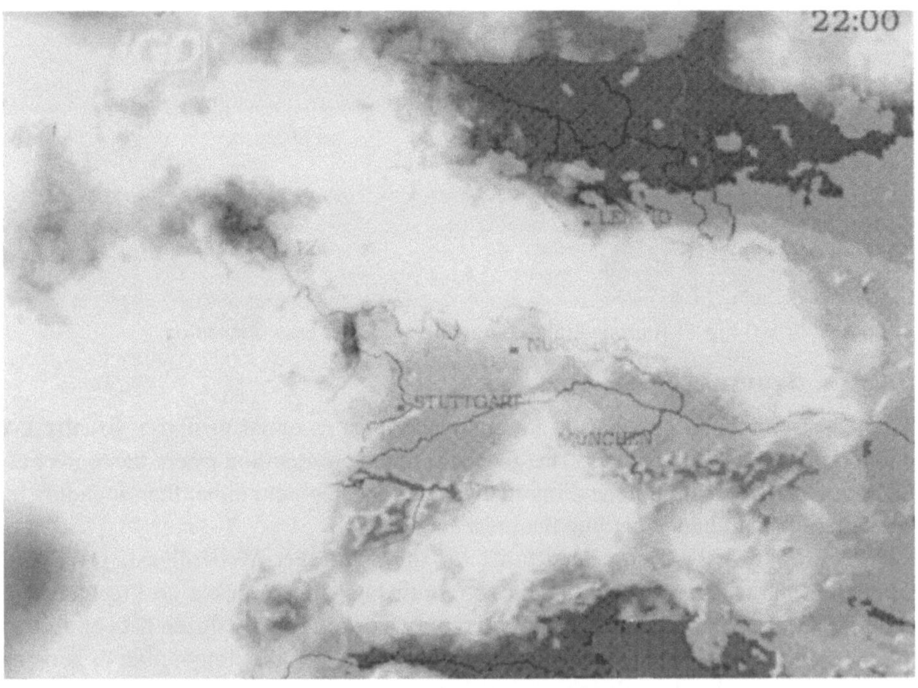

Fig. 3. Fractal clouds with coloring

So we chose fractal functions (Rescale And Add method) to generate the cloud objects we then model according to the simulation data. During this visual processing we always preserve the mean values of the forecast for each area. And the very local error we make with the fractal variation is negligible compared to the medium error of the forecast computation itself.

The results are very promising. Several national and local television stations already broadcast our video sequences we call "pseudo satellite films" in their daily program. It is now possible to visualize cloud development intuitively understandable and, thus, perceptually more effective for lay audience.

2.3 Cloud Coloring

Apart from presenting clouds that look like clouds by using fractal functions, we color the clouds according to their meteorological activity. This value is extracted from the simulation data for each point and gives information about the probability of precipitation and its expected strength.

Now we show the clouds as seen from above looking downwards onto the ground, but we color them as they would appear when seen from the ground looking upwards. So more "dangerous" clouds look dark grey, and white colors are assigned to more "friendly" clouds.

This seems paradoxical, but by doing this we use the individual's every day experience of weather phenomena. Everybody is used to stand on the ground and look into the sky. White clouds are always friendly, and dark ones appear rather threatening. This coloring is easily understandable to most viewers. If we colored the clouds correctly as they would look from above according to their dust content, height, refraction values, and other parameters, this would mean nothing to most individuals.

Lightning areas are presented as flashing white areas in the image in the parts where lightning and thunder are predicted. This is also a visualization method that is easy to understand by non-expert viewers.

2.4 Temperature Colormap

Apart from cloud constellation and movement, lay audiences are always most interested in the temperature development. The goal was to also present this information in an intuitively understandable way. The predicted temperature values come on the same coarse grid out of the simulation as the cloud-specific data. But here smoothing by interpolation is sufficient. People are used to the fact that temperature always changes rather gradually. But for clearer presentation iso-areas can also be used. And, since temperature can only very seldom be visually perceived in real life, the viewers do not really expect realistic presentations on the television screen, as it is the case with clouds.

Here the colors used to represent the different temperatures are most important. Several aspects must be considered to find a perceptually effective way to map the values to colors. Again the individual's every day experience should be the basis of the coloring method. Almost everywhere people are used to associating blue with cold and red with hot temperatures. Also the color green is always associated with a moderate "O.K." value. White can be used as a "neutral" value with no strong effect. This is why we use a colormap that ranges from blue (very cold) over green or white (medium temperature) and yellow-orange (rather warm) to red (very hot). A colormap that has purple and red at the extreme ends is not suitable, since these colors are very close but do represent extremely opposite values. White has the advantage that it is mostly not regarded as a color itself and does not distract the viewer from important values with their colorings.

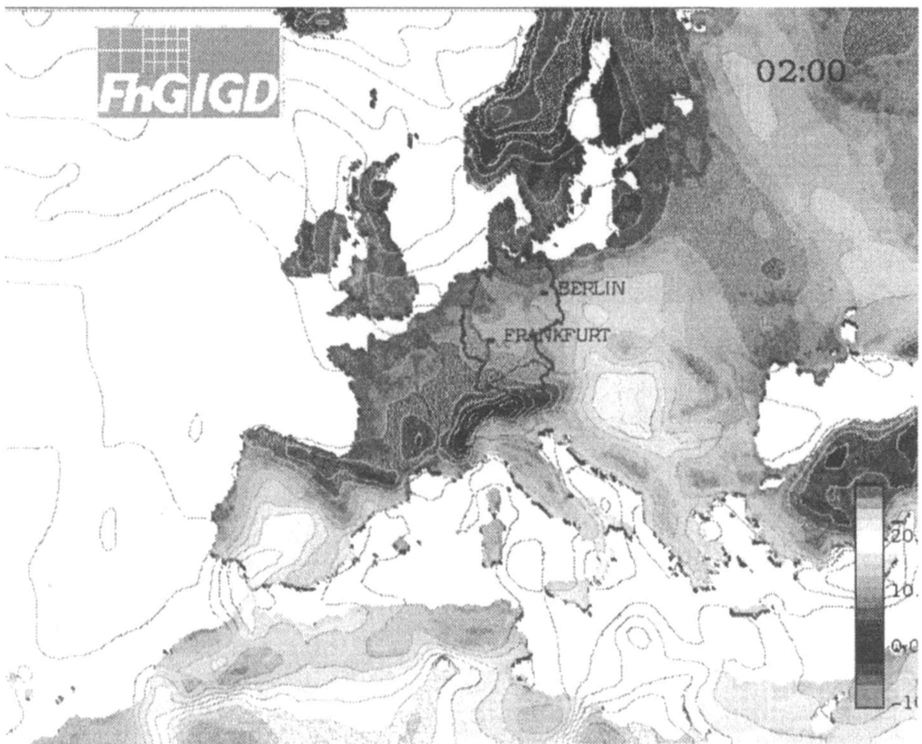

Fig. 4. Temperature over continental Europe

When visualizing ozone concentration as another example, red can be used for high values using the association with a warning, while green can be assigned to low concentrations representing well-being and life.

An important question here is whether the coloring should be dynamic. Should we assign different temperature values to the colors in different seasons of the year? Should blue represent 20 degrees Celsius below freezing in winter and 5 degrees above freezing in summer, because both values are relatively "very cold" in their season? Or does this confuse the viewer and does he or she expect a static colormap, where blue always stands for 20 degrees below freezing for example?

One argument often used against a multispectral colormap is the fact that it has the effect of segmentation on the viewer's side even when it actually represents continuous scalar values with no segmentation whatsoever. However, when visualizing temperature, e.g., for example for lay audience, this can often be a desired effect.

Viewers can look at the screen and say: "Look, we live in the yellowish colored area. That means we can go swimming tomorrow!" Such a meaning can be assigned to different color segments and can be extracted as different meanings by the viewer as "too cold", "fairly mild", "just right for swimming", or "too hot".

2.5 Dynamics

One of the most important aspects of the weather is its dynamics, and television spectators are very interested in getting this information. "Does this cold front move over us during Saturday morning or does it move slowly and effect us the whole weekend?"

Predictions calculated with the European Model at the German meteorological office simulate a weather situation every hour. We show the weather dynamics by visualizing every discrete time step and showing the results in an animated sequence. This has proven to be very successful, since the viewer sees the day going by (watching the clock displayed in each image) and the clouds or temperature areas move relatively to time.

Even though people are used to seeing the clouds move smoothly in reality, they accept our not very smooth animation (one dataset and image per hour), because they have been trained to watch and interpret satellite films (two images per hour) in the past years of TV weather forecasting.

2.6 Icons

Since many people are very used to gain information about the weather through icons presenting suns, clouds, snow flakes or other symbols, we have also integrated these symbols in our system. By superimposing these icons upon other visualized results, additional meaning can be added. Tests must now show how much information can be combined effectively in one image.

2.7 Vector Data

Also research is currently carried out in our group about intuitively understandable forms of presenting vector fields with wind data. We will try particles like leaves being blown over the forecast area, streamlines, or arrow symbols. Also flags, vanes, or wind bags at static locations reacting to the wind fields could be appropriate presentations.

2.8 Three-Dimensional Presentation

By rendering three-dimensional terrain data and visualizing 3D meteorological data, we will be able to better present local weather phenomena. Our current work is therefore directed towards such a system. It will allow a presentation like "What would we see tomorrow if we stood on that mountain and looked around?". Also, it will support arbitrary flights and views over the prediction area and through the clouds. We have already developed a first 2 1/2 dimensional prototype which is installed at the DWD for testing.

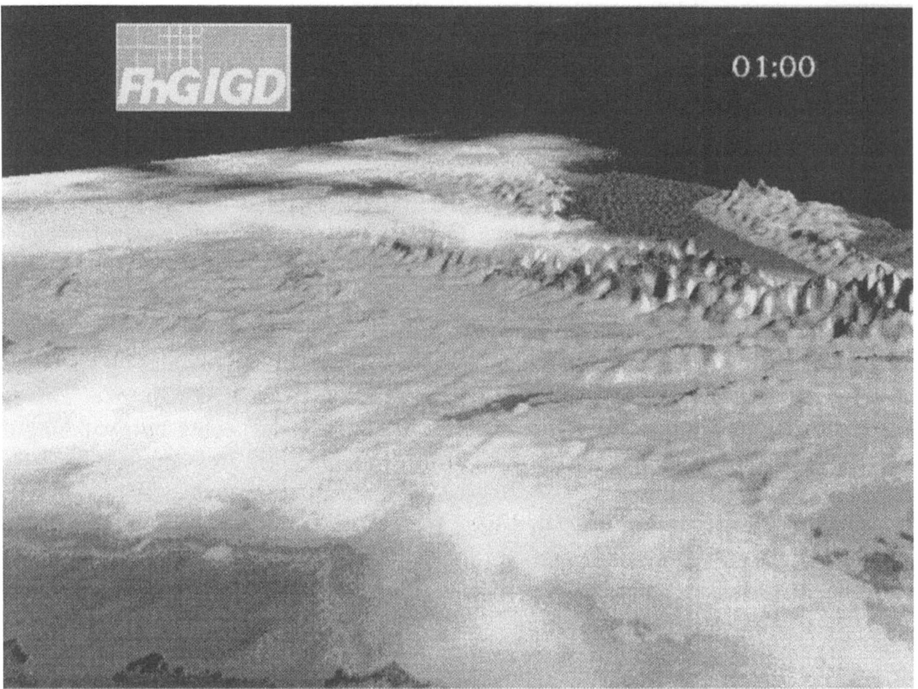

Fig. 5. Fractal clouds over three-dimensional polygonal terrain. View from above England in south-east direction

3 Research Aspects in the Field of Perception

Some areas of our current research are especially strongly related to the field of perception. Among them are the following two areas:

3.1 Maximum Information Content

To what extent can we keep on putting more and more information into one image? Currently the elevation of the ground is colored according to its height. Above this there are clouds with varying opacities and colors giving information about cloud coverage and meteorological activity. Also colored areas representing temperatures can be drawn with a constant transparency value above the height image. In addition to that we have a digital clock showing the time of day in a corner of the screen. All of this information can be perceived quite well by the non-expert viewers. But can we add more information to a single animated sequence?

Our system is capable of showing meteorological symbols like fronts, icons, or text strings as overlay. Also several values could be visualized together (e.g., clouds and temperature). And can we also show the type and amount of precipitation in the same sequence where we show the clouds?

Here exact tests are necessary before we start overloading our presentations with information. But it is also a question how many separate sequences showing different values develop in the same time interval can be combined by the non-expert viewer.

3.2   Audience's Ability to Learn

A very interesting field to investigate is the way audience learn to interpret presentations correctly. How did the television audience from today learn to interpret satellite images showing the weather development of the past? Is a learning in this respect possible at all? If so, can we find a way of visualizing meteorological data perceptually most efficient that the audience must get used to? Can we gradually change the way of presenting such information and continuously teach the television spectators to interpret it correctly? How much can a non-expert audience learn in that aspect?

Literature:

S. Arndt, T. Frühauf, K. Karlsson, F. Schröder; "Integration of Compute Servers in an Environment for Distributed Simulation and Visualization", ECUC'92 Proceedings, European Convex Users Conference, 1992

D. S. Dyer; "A Dataflow Toolkit for Visualization", IEEE Computer Graphics and Applications; July Issue 1990

J. L. Encarnação, W. Felger, M. Frühauf, M. Göbel, K. Karlsson; "Interactive modeling in high performance scientific visualization - the Vis-a-Vis project"; Computers in Industry, North-Holland, Volume 19, No. 2, 1992

W. Felger, F. Schröder; "The Visualization Input Pipeline - Enabling Semantic Interaction in Scientific Visualization", EUROGRAPHICS'92 Proceedings; Computer Graphics Forum; NCC Blackwell Publishers, 1992

G. Sakas, F. Schröder, H. J. Koppert; "Pseudo-Satellitefilm - Using Fractals to Enhance Animated Weather Forecasting"; EUROGRAPHICS'93 Proceedings; Computer Graphics Forum; NCC Blackwell Publishers, 1993

F. Schröder; "Visualizing Meteorological Data for a Lay Audience" IEEE Computer Graphics and Applications; Sept. Issue 1993

If you have any concerns about our products,
you can contact us on
ProductSafety@springernature.com

In case Publisher is established outside the EU,
the EU authorized representative is:
Springer Nature Customer Service Center GmbH
Europaplatz 3, 69115 Heidelberg, Germany

Printed by Libri Plureos GmbH
in Hamburg, Germany